Programmed Visions

Software Studies

Matthew Fuller, Lev Manovich, and Noah Wardrip-Fruin, editors

Expressive Processing: Digital Fictions, Computer Games, and Software Studies
Noah Wardrip-Fruin, 2009

Code/Space: Software and Everyday Life
Rob Kitchin and Martin Dodge, 2011

Programmed Visions: Software and Memory
Wendy Hui Kyong Chun, 2011

Programmed Visions

Software and Memory

Wendy Hui Kyong Chun

The MIT Press
Cambridge, Massachusetts
London, England

First MIT Press paperback edition, 2013

© 2011 Massachusetts Institute of Technology

This book was set in Stone Sans and Stone Serif by Toppan Best-set Premedia Limited.

Library of Congress Cataloging-in-Publication Data

Chun, Wendy Hui Kyong, 1969–
Programmed visions : software and memory / Wendy Hui Kyong Chun.
   p.   cm. — (Software studies)
Includes bibliographical references and index.
ISBN 978-0-262-01542-4 (hc. : alk. paper)—978-0-262-51851-2 (pb. : alk. paper)
1. Computer software—Development—Social aspects. 2. Software architecture—Social aspects. 3. Computer software—Human factors. I. Title.
QA76.76.D47C565   2011
005.1—dc22

                                                                                    2010036044

Contents

Series Foreword   vii

Preface: Programming the Bleeding Edge of Obsolescence   xi

Introduction: Software, a Supersensible Sensible Thing   1

You   13

I   **Invisibly Visible, Visibly Invisible   15**

1   **On Sourcery and Source Codes   19**

Computers that Roar   55

2   **Daemonic Interfaces, Empowering Obfuscations   59**

II   **Regenerating Archives   97**

3   **Order from Order, or Life According to Software   101**

The Undead of Information   133

4   **Always Already There, or Software as Memory   137**

Conclusion: In Medias Res   175

Epilogue: In Medias Race   179

You, Again   181

Notes   183

Index   233

Series Foreword

Software studies aims to find ways of expanding and intensifying reflection on software and computational culture in general. The problems it works on are rather unavoidable since software, and the underlying ideas and techniques that it embodies, is a crucial, if underacknowledged, element of everyday life. Few parts of human culture remain untouched by software, but there are relatively fewer means by which to evaluate it. The Software Studies book series aims to contribute to a certain balancing out of this ratio.

The ability to understand its preconditions and basal factors is in turn essential for any field of endeavor to prosper and to renew itself. To ally such an understanding with a synthetic approach, which brings together some of the iterations of a foundational set of ideas as they move through different fields and are changed by them as they in turn change those that they provide new insights to, is crucial. As this book shows, the ideas of code and of programmability underlie software. In turn, they form a set of idioms and techniques to shape and make possible other areas of life.

While *Programmed Visions* operates as a sustained introduction to the ideas of software, code, and programmability as they work in relation to computation, the book is also a meditation on how this model proliferates, by various means, into systems such as living materials that are in turn understood to be bearers of a form of code that instructs their growth and that can, by further convolution, be read as a print out of the truth of an organism. Indeed, Chun's book shows how, in nuanced and intriguing ways, the idea of code in biology anticipates that in computing. Thus, the idea of programmability proliferates into other pasts.

That computing is something that has a history is, three generations removed from the first electronic computers, relatively well established. The study of that history itself has grown from a focus on canonical surveys and detailed and vivid oral histories to a very fruitful proliferation of focuses, problematics, and methodological scope. *Programmed Visions* places the field of software studies in direct dialogue with that of computing history, but it also suggests that in order to work through history, we need to be able to bring other scales into account, from feelings to geopolitics and the conceptual and ideological orderings that are operative in them.

One of the operations evaluated here is the idea that one thing can stand in for, or be seen as equivalent to, another. This is the essential idea of a code. Systems of equivalence and codification such as capitalism, the universal Turing machine, management, structuralism, each has its own idiosyncrasies, and each, as constructive systems, has its own capacities for invention. Chun's claim, in an interlude text in this book, is that the computer, and software in particular, has gone one step further, becoming a metaphor for metaphor, a means by which other metaphors are filtered and arranged, becoming in turn a system of universal experiential machining. This is one reason the computer cannot be written off, or lauded, as a simply crazily rationalist machine. There is a velocity, idiosyncrasy, and thickness to the changes wrought by software that makes it a fundamentally tricky phenomenon, potentially rich rather than inherently reductive, but not automatically so.

One other set of phenomena that these qualities couple with are the means of assigning value to things. The degradation experienced in the neoliberal moment is partly in the abstractions it operates by: that relations, singular qualities of inherence in the world, are exchanged for equivalences; that money becomes the secret means by which a table may be transmogrified into a meal and a house may be turned into a debt. In the secret ironic engine undergirding economics, equivalences are exchanged for sames. These sames may be goods, the same dull coffee places in cities across the overdeveloped world, the same infrastructure of contracts, law, and possession, and the same operating systems that accompany them.

The ability of numbers, statements, currencies, or other signs to stand in for all kinds of things gives systems of abstraction and generalization immense power, especially when they can be made to line up into larger-scale structures, producing veritable machines. *Programmed Visions* gives us a means of understanding such processes, but also importantly understanding how software is the code that works to disintermediate these systems. Thus, to understand the contemporary situation, it is not enough solely to recognize the operations of the economy, or even to be able to interrogate the morphological expressivity of a genetic array, but also to understand the very mechanisms that conjoin them. And here, software's capacity to handle relations, equivalences,and sames is also something that, as well as bearing the capacity for indefensible reductions, also makes it deeply productive. Software, in its relations with other things, brings a capacity of synthesis to multiple scales of reality, acting as a condition of thought, of imagination, investing them with multiple kinds of motility and conjunction. In turn, one of the imaginaries that invests this synthetic domain is a technocratic dreamwork of understanding, interpretability, ambivalent optimization, but also of instrumentalisation and restructuration running in a recursive mode that reinforces systems of sames.

This is a necessarily complicated, highly intriguing, series of transitions and the elegance with which Chun marks these moments of the waxing and waning of integrations and encodings is testament both to the expressivity of the systems that she interprets and to the skill with which her analyses are assembled. The very dynamic range of the materials that the book discusses indeed compels what the book both celebrates and exemplifies: a means of thinking "in the middle of things." This feel for both immanence and abstraction drives *Programmed Visions* in its figuring out of the relations between the different loci that it inhabits, and it is one that is marked by multiple resonances of vicissitude and pleasure. It is in these transitions too that the book engenders its relation to memory, the regenerative capacity that is needed when one does not have an absolute overview. Memory allows us to see patterns, to unlock codes, even in a world of ongoing change. *Programmed Visions* sets such a capacity in contrast to the figure of memory as simple storage, or "hardened" information, and offers a new reading of the relationship between them. In broader terms, the book commends us to keep looking at what becomes soft, that which ossifies or proliferates by staying the same, what multiplies and what grows anew. With an urgency that cannot be rushed, we are here presented with the materials to carry out such work.

Matthew Fuller

Preface: Programming the Bleeding Edge of Obsolescence

This book was inspired by the many lives of new media—by the ways that it not only survives, but also thrives on, cycles of obsolescence and renewal.

In the early 2000s, new media seemed to be dead, and the utopian and dystopian discourses around the World Wide Web and Y2K were exposed for what they were: hype. Gone were the celebrations of the "new economy," virtual reality, and cyberspace. The term *new media* even seemed "old": the New York New Media Association folded in 2003, and many New Media Groups within corporate structures (Apple, Gannett, etc.), and many new media companies disappeared.[1] Everyone was on the Internet—new media was everywhere—but new media seemed boring; the reality of surfing the net did not compare to the glitzy cyberpunk visions touted by *Mondo 2000*.

By 2008, however, the future was, once more, in fashion, and there was a growing impatience with the so-called critical hindsight that flourished after the dot.bombs and 9/11. Rather than sobering if banal reassessments of the Internet as a "double-edged sword" that aids both terrorists and victims, the main strain of both popular and scholarly new media analysis stressed future possibilities and sought to outline the next big thing: mobile mobs, Web 3.0, cloud computing, radical topsight, and so on. A sense that something had really changed, as well as a desire to capitalize on this change, fueled this renewal: the returns of new media are linked to the promise of financial returns. Silicon Valley, if not Alley, had recovered from the demise of the "new economy"; Google was everywhere in every possible form; iPhones and BlackBerries had proliferated; even Granny was on Facebook.com. Every social movement, every social protest appeared to be wired; newspaper companies were folding and television stations laid off staff as content migrated online; everyone, it appeared, was bombarding one another with 140-character-long tweets, and no one seemed to care.

This future 2.0, like Web 2.0 or 3.0, was not as utopian or as bold as its mid-1990s predecessor, *the* future. No one was prophesying the end of all brick-and-mortar businesses; there were no upbeat yet paranoid commercials promising the end to racial discrimination and the beginnings of a happy global village; there were no must-read cyberpunk novels or films outlining cyberspace's gritty, all-encompassing nature, although *new media* does now encompass the bio- and nanotech. Instead, even within this optimism, there was a dim yet gnawing sense that this too will pass, that every next big thing is also the next big bubble (if it's anything at all). To call something new, after all, is to guarantee its obsolescence, and this hopeful return to the future as future simple—as what will be, as what you will do, as a programmed upgrade to your already existing platform—constantly recedes and disappears. Although this cycle of the ever-returning and ever-receding new mirrors the economic cycle it facilitates, the undeadness of new media is not a simple consequence of economics; rather, this book argues, this cycle is also related to new media's (undead) logic of programmability. New media proliferates "programmed visions," which seek to shape and to predict—indeed to embody—a future based on past data.

This book addresses this concept of programmability through the surprising materialization of software as a *thing* in its own right. It argues that the hardening of programming into software and of memory into storage is key to understanding new media as a constantly inspiring yet disappointing medium of the future. It links this hardening to several factors: computing's gendered and military history, foundational parallels between the fields of genetics and computing technology, long-standing visions of a stable archive of knowledge as driving human progress, and a general, neoliberal trend to personalize power (to make power touch each and all). All this has made the computer, understood as networked software and hardware machines, both an instrument and a symptom of neoliberal governmental power. It has made it an instrument of both causal pleasure and extreme frustration, a means of navigation and obfuscation.

This book, however, does not seek to condemn computers as simple neoliberal tools or to view user empowerment as a form of imprisonment. Computers are mediums of power in the fullest senses of both words. Through them, we can pleasurably create visions that go elsewhere, specters that reveal the limitations and possibilities of user and programmer, choices that show how we can rework neoliberal formulations of freedom and flexibility. Specters haunt us through our interfaces—by working with them we can collectively negotiate the dangers and pleasures of the worlds they encapsulate and explode.

Acknowledgments

I am very grateful to all those who have read and sponsored various parts of this book. I owe special thanks to Matthew Fuller and Florian Cramer who read drafts of the whole book, and to Lisa Gannett, N. Katherine Hayles, Adrian Mackenzie, the editors of *Grey Room*, the editorial board of *Critical Inquiry*, and the Critical Code Studies Working Group, who all read and offered critiques of portions of it. Their comments have immeasurably improved this book. I have learned much and received great support from my colleagues in the Department of Modern Culture and Media. To Chris Csikszentmihalyi, Arindam Dutta, Liz Canner, Lynn Festa, Thomas Keenan, and Mary Ann Doane, I owe much inspiration and good cheer. I am also grateful to my incredible research assistant Ioana Jucan for her impeccable work and to Robin Davis for her assistance with the images. To the fantastic editorial machine at MIT—Doug Sery, Katie Helke, and Kathleen Caruso—I owe an enormous thanks. Without the love and support of my sweetie Paul Moorcroft, this book would not have been possible.

Research for this book was supported by grants, fellowships and leave from Brown University (in particular, a Henry Merritt Wriston Fellowship and a Edwin and Shirley Seave Faculty Fellowship from the Pembroke Center for Teaching and Research on Women)—I am grateful to Brown University for its financial and academic support. A fellowship from the Radcliffe Institute for Advanced Study was crucial to conceiving the manuscript, and a travel award from the Lemelson Center at the National Museum of American History, Smithsonian Institute made it possible for me to do archival work. I began writing in earnest while a visiting scholar in Harvard University's History of Science Department—I would like to thank Harvard and my hosts for their invaluable support.

Fragments of the book have been published in *Configurations*, *Grey Room*, and *Critical Inquiry*.

Boston, Massachusetts
August 2010

Introduction: Software, a Supersensible Sensible Thing

Debates over new media resonate with the parable of the six blind men and the elephant. Each man seizes a portion of the animal and offers a different analogy: the elephant is like a wall, a spear, a snake, a tree, a palm, a rope. Refusing to back down from their positions since they are based on personal experience, the wise men engage in an unending dispute with each "in his own opinion / Exceeding stiff and strong / Though each was partly in the right, / And all were in the wrong!" The moral, according to John Godfrey Saxe's version of this tale, is: "So oft in theologic wars, / The disputants, I ween, / Rail on in utter ignorance / Of what each other mean, / And prate about an Elephant / Not one of them has seen!"[1] It is perhaps irreverent to compare a poem on the incomprehensibility of the divine to arguments over new media, but the invisibility, ubiquity, and alleged power of new media (and technology more generally) lend themselves to this analogy. It seems impossible to know the extent, content, and effects of new media. Who can touch the entire contents of the World Wide Web or know the real size of the Internet or of mobile networks? Who can read and examine all time-based online interactions? Who can expertly move from analyzing social networking sites to Japanese cell phone novels to hardware algorithms to databases? Is a global picture of new media possible?

In response to these difficulties, many within the field of new media studies have moved away from specific content and technologies toward what seems to be common to all new media objects and moments: software. All new media objects allegedly rely on—or, most strongly, can be reduced to—software, a visibly invisible or invisibly visible essence. Software seems to allow one to grasp the entire elephant because it is the invisible whole that generates the sensuous parts. Based on and yet exceeding our sense of touch—based on our ability to manipulate virtual objects we cannot entirely see—it is a magical source that promises to bring together the fractured field of new media studies and to encapsulate the difference this field makes. To know software has become a form of enlightenment: a Kantian release from self-incurred tutelage.

This notion of knowing software as a form of enlightenment—as a way to comprehend an invisible yet powerful whole—is not limited to the field of new media studies. Based on metaphor, software has become a metaphor for the mind, for culture, for ideology, for biology, and for the economy. Cognitive science, as Paul Edwards has shown, initially comprehended the brain/mind in terms of hardware/software.[2] Molecular biology conceives of DNA as a series of genetic "programs." More broadly, culture itself has been posited as "software," in opposition to nature, which is "hardware."[3] Although technologies, such as clocks and steam engines, have historically been used metaphorically to conceptualize our bodies and culture, software is unique in its status as metaphor for metaphor itself. As a universal imitator/machine, it encapsulates a logic of general substitutability: a logic of ordering and creative, animating disordering. Joseph Weizenbaum has argued that computers have become metaphors for all "effective procedures," that is, for anything that can be solved in a prescribed number of steps, such as gene expression and clerical work.[4]

The clarity offered by software as metaphor—and the empowerment allegedly offered to us who know software—however, should make us pause, because software also engenders a sense of profound ignorance. Software is extremely difficult to comprehend. Who really knows what lurks behind our smiling interfaces, behind the objects we click and manipulate? Who completely understands what one's computer is actually doing at any given moment? Software as metaphor for metaphor troubles the usual functioning of metaphor, that is, the clarification of an unknown concept through a known one. For, if software illuminates an unknown, it does so through an unknowable (software). This paradox—this drive to grasp what we do not know through what we do not entirely understand—this book argues, does not undermine, but rather grounds software's appeal. Its combination of what can be seen and not seen, can be known and not known—its separation of interface from algorithm, of software from hardware—makes it a powerful metaphor for everything we believe is invisible yet generates visible effects, from genetics to the invisible hand of the market, from ideology to culture.

Every use entails an act of faith, and this book tries to understand what makes this trust possible not in order to condemn and move "beyond" computer software and interfaces, but rather to understand how this combination of visibility and invisibility, of past experiences with future expectation, makes new media such a powerful thing for each and all. It also takes seriously new media's modes of repetition and transmission in order to understand how they open up gaps for a future beyond predictions based on the past. Computers—understood as software and hardware machines—this book argues, are mediums of power. This is not only because they create empowered users, but also and most importantly, because software's vapory materialization and its ghostly interfaces embody—conceptually, metaphorically, virtually—a way to navigate our increasingly complex world.

How Soft Is Software?

Software is, or should be, a notoriously difficult concept. Historically unforeseen, barely a thing, software's ghostly presence produces and defies apprehension, allowing us to grasp the world through its ungraspable mediation.

Computer scientist Manfred Broy describes software as "almost intangible, generally invisible, complex, vast and difficult to comprehend." Because software is "complex, error-prone and difficult to visualize," Broy argues, many of its "pioneers" have sought to make "software easier to visualize and understand, and to represent the phenomena encountered in software development in models that make the often implicit and intangible software engineering tasks explicit."[5] Software challenges our understanding not only because it works invisibly, but also because it is fundamentally ephemeral—it cannot be reduced to program data stored on a hard disk. Historian Michael Mahoney describes software as "elusively intangible. In essence, it is the behavior of the machines when running. It is what converts their architecture to action, and it is constructed with action in mind; the programmer aims to make something happen."[6] Consequently, software is notoriously difficult to study historically: most "archived" software programs can no longer be executed, and thus experienced, since the operating systems and machines, with which they merge when running, have disappeared. Although these systems can be emulated, what is experienced is a reconstruction.[7] Hence, not only does software's ephemerality make analysis difficult, so does the lack of clear boundaries between running programs and between running software and live hardware. Theorist Adrian Mackenzie aptly calls software a "neighbourhood of relations"; "in code and coding," he argues, "relations are assembled, dismantled, bundled and dispersed within and across contexts."[8] Software "pioneers" Herman H. Goldstine and John von Neumann, in their 1940s explication of programming, similarly described it as "the technique of providing a dynamic background to control the automatic evolution of a meaning."[9]

To be apprehended, software's dynamic porousness is often conceptually transformed into well-defined layers. Software's temporality, in other words, is converted in part to spatiality, process in time conceived in terms of a process in space. Historian Paul Ceruzzi likens software to an onion, "with many distinct layers of software over a hardware core."[10] Application on top of operating system, on top of device drivers, and so on all the way down to voltage charges in transistors. What, however, is the difference between an onion's layers and its core? Media archeologist Friedrich Kittler, taking this embedded and embedding logic to its limit, has infamously declared "there is no software," for everything, in the end, reduces to voltage differences. More precisely, he contends, "there would be no software if computer systems were not surrounded . . . by an environment of everyday languages. This environment . . . since a famous and twofold Greek invention, consists of letters and coins, of books and bucks."[11] Less controversially, Mahoney has argued that software "is an artifact of computing in the business and government sectors during the '50s"; software, as Paul Ceruzzi and Wolfgang Hagen have shown, was not foreseen: the engineers building high-speed calculators in the mid-1940s did not plan or see the need for software.[12]

At first, software encompassed everything that was not hardware, such as services. The term *soft*, as this book elaborates, is gendered. Grace Murray Hopper claims that the term *software* was introduced to describe compilers, which she initially called "layettes" for computers; J. Chuan Chu, one of the hardware engineers for the ENIAC, the first working electronic digital computer, called software the "daughter" of Frankenstein (hardware being the son).[13] Software, as a service, was initially priced in terms of labor cost per instruction.[14] Herbert D. Benington remarks that attendees at the 1956 symposium on advanced programming methods for digital computers were horrified that his Lincoln Laboratory group, working on what would become the groundbreaking SAGE (Semi-Automatic Ground Environment) Air Defense System, could do no better than $50 per instruction. In that 1956 address Benington also stresses the growing importance of software: "our colleagues who build computers," he notes, "have come to realize that a computer is not useful until it has been programmed."[15] As this statement reveals, the word *program*, at that time, was predominantly a verb, not a noun.[16]

Legal battles over software copyrights and patents make clear the stakes of this transformation of software from a service, priced per instruction, to a thing. Not surprisingly, software initially was considered neither patentable nor copyrightable because of its functional, intangible, and "natural" status. The U.S. Supreme Court in 1972 first rejected engineers Gary Benson and Arthur Tabbot's claim to patent an algorithm for converting digital into binary digits. It decided, as legal scholar Pamela Samuelson argues, that "mathematical innovations should be treated like scientific truths and laws of nature, and scientific truths and laws of nature are unpatentable subject matter."[17] Software algorithms, in other words, were "natural" mental processes, not artificial things. As Samuelson and as legal scholar Margaret Jane Radin both note, key to the eventual patenting of software was its transformation from a set of instructions to a machine.[18] In 1981, the Supreme Court in *Diamond v. Diehr, 450 U.S. 175* (1981) upheld the patenting of an algorithmic-based process for curing rubber because the algorithm resulted in a tangible physical process: it cured rubber. By 1994, the U.S. Court of Appeals Federal Circuit held in *In re Alappat* (1994) that all software was inherently machinic, since it changed the material nature of a computer: "a general purpose computer in effect becomes a special purpose computer once it is programmed to perform particular functions pursuant to instructions from program software."[19] A change in memory, it seems, a change in machine.

As a physical process, however, software would seem uncopyrightable.[20] Copyright
seeks to protect creative expression; as Radin notes, patents and copyrights were supposed to be mutually exclusive: "Copyright is supposed to exclude works that are
functional; patent is supposed to focus on functionality and exclude texts."[21] To
address this contradiction, the U.S. Congress changed the law in 1975, so that expressions, as opposed to the actual processes or methods, adopted by the programmer
became copyrightable.[22] The difference, however, between expression and methods
has been difficult to determine, especially since the expression of software has not
been limited to source code.

Further, copyright law insists on the tangibility of the copy, where a copy is a "fixation in a tangible medium of expression." Performances thus were initially considered
to be outside the purview of copyright.[23] Although information is often considered to
be immaterial, the forces behind copyrighting (and taxing) software stress the fact
that, regardless of information's ephemerality, information is always embodied; it
always, as Matthew Kirschenbaum argues, leaves a trace.[24] Indeed, digital information
has divorced tangibility from permanence, with "courts and commentators in the
United States adopt[ing] the notion that the momentary arrangement of electrons in
a computer memory, which we might have thought of as intangible information,
amounts to a tangible physical object, a copy."[25] Since, as I have argued elsewhere,
computer reading is a writing elsewhere, viewing the momentary arrangement of
electrons in memory as a tangible copy technically makes all computer reading a
copyright infringement. Indeed, this redefinition of copy as thing, as Radin notes, has
had far-reaching consequences since "a great many activities that were not covered
by copyright in the offline environment are being brought under copyright—that is,
under control of an owner—in the online environment. . . . The physical analogy to
browsing in a bookstore is obliterated by the more powerful assimilation of the activity
involved in a physical object—the production of physical 'copies' by a computer."[26]
This definition also muddies questions of responsibility: given that every networked
computer regularly downloads all materials in a network and then erases those not
directly addressed to it, should everyone whose computer has unwittingly downloaded
child pornography or pirated media be prosecuted?

These changes, brought about by the "hardening" of software as textual or machinic
thing through memory, point toward a profound change in our understanding of what
is internal and external, subject and object. According to Radin, "the distinction
between tangible objects and intangible information is a distinction upon which
much of our modern understanding of the world was built, and hence, from which a
great many legal categorizations derive," for this traditional distinction "owes much
to the 'modernist' dichotomies of the Enlightenment—between subject and object,
between autonomous persons and heteronomous things."[27] The notion of intellectual
property, which seems to break this dichotomy, was initially a compromise, she

contends, between the Enlightenment notion that the intellect was internal and property external.[28] (It is not simply, though, that information was once inside a person and then externalized, but also that information was considered inseparable from a person. Symptomatically, the meaning of information has moved from "the action of informing . . . the formation or moulding of the mind or character, training, instruction, teaching" to "knowledge communicated concerning some particular fact, subject, or event."[29]) Crucially, Radin argues that the information age has compromised the compromise that intellectual property represents, since, by breaking down the distinction between tangibility and intangibility, it conceives of information, whether internal or external, as always external to the self (hence the patentability of genes). As I've argued elsewhere, the Internet and computers—which have offered enlightenment for all—have exploded enlightenment by literalizing it.

Software as thing has led to all "information" as thing. Software as thing reconceptualizes society, bodies, and memories in ways that both compromise and extend the subject, the user. Importantly, software as thing cannot be reduced to software as a commodity: software as "thing" is a return to older definitions of thing as a "gathering," as pertaining to anything related to "man."[30] Treating software as a thing means treating it, again, as a neighborhood, as an amalgamation. It also means thinking through its simultaneous ambiguity and specificity. Further, it means thinking beyond this legal history, this legal framework, toward the historical and theoretical stakes of the reemergence of things as relations. Indeed, this book argues that the remarkable process by which software was transformed from a service in time to a product, the hardening of relations into a thing, the externalization of information from the self, coincides with and embodies larger changes within what Michel Foucault has called *governmentality*. Software as thing is a response to and product of changing relations between subjects and objects, of challenges brought about by computing as a neoliberal governmental technology.

Soft Government

According to Foucault, governmentality and government broadly encompass acts and institutions that govern, or steer, conduct and thus cannot be reduced to the state. (Not coincidentally, the term *cybernetics* is derived from the Greek term "kybernete" for governing.) As Colin Gordon notes, government for Foucault is "the conduct of conduct," that is, "a form of activity aiming to shape, guide or affect the conduct of some person or persons." Governmentality could concern "the relation between self and self, private interpersonal relations involving some form of control or guidance, relations within social institutions and communities and, finally, relations concerned with the exercise of political sovereignty."[31] The move from the Enlightenment, with its dichotomy of subjects and objects, to our current compromised situation corresponds to a transition from liberal to neoliberal governmentality (and, even further, to a neoconservative one).

Liberal governmentality, which emerged during the eighteenth century, is an "economic government": government that embraces both liberal political economy and the principle of noninterference. It is based on two principles: the principle of blind self-interest and the principle of freedom. According to its vision, actors, who cannot know the whole picture, blindly and freely follow their own self-interests so that "the invisible hand of the market" can magically incorporate their actions into a system that benefits all. This unknowability is fundamental, for it enables a transition from sovereign to liberal forms of governmentality. The liberal market undermines the power of the monarch by undermining his or her knowledge: no one can have a totalizing view. It also consumes freedom: it both produces freedom and seeks to control it.[32] Liberal governmentality also makes possible biopolitical power: a collection of institutions and actions focused on "taking care" of a population, rather than a territory, focused on masses rather than on sovereign subjects.

Historically, computers, human and mechanical, have been central to the management and creation of populations, political economy, and apparatuses of security.[33] Without them, there could be no statistical analysis of populations: from the processing of censuses to bioinformatics, from surveys that drive consumer desire to social security databases. Without them, there would be no government, no corporations, no schools, no global marketplace, or, at the very least, they would be difficult to operate. Tellingly, the beginnings of IBM as a corporation—the Herman Hollerith's *Tabulating Machine Company*—dovetails with the mechanical analysis of the U.S. census.[34] Before the adoption of these machines in 1890, the U.S. government had been struggling to analyze the data produced by the decennial census (the 1880 census taking seven years to process). Crucially, Hollerith's punch-card-based mechanical analysis was inspired by the "punch photograph" used by train conductors to verify passengers.[35] Similarly, the Jacquard Loom, a machine central to the industrial revolution, inspired (via Charles Babbage's "engines") the cards used by the Mark 1, an early electromechanical computer. Scientific projects linked to governmentality also drove the development of data analysis: eugenics projects that demanded vast statistical analyses, nuclear weapons that depended on solving difficult partial differential equations.[36]

Importantly, though, computers in the period this book focuses on (post–World War II) coincide with the emergence of neoliberalism. As well as control of "masses," computers have been central to processes of individualization or personalization. Neoliberalism, according to David Harvey is "a theory of political economic practices that proposes that human well-being can best be advanced by liberating individual entrepreneurial freedoms and skills within an institutional framework characterized by strong private property rights, free markets, free trade."[37] Although neoliberals, such as the Chicago School economist Milton Friedman, claim merely to be resuscitating classical liberal economic theory, Foucault argues that neoliberalism differs from liberalism in its stance that the market should be "the principle, form, and model for a state."[38] It contends that individual economic and political freedom are tied together: competitive capitalism, Friedman writes, "is a system of economic freedom and a necessary condition for political freedom."[39] Harvey argues that neoliberalism has thrived by creating a general "culture of consent"—even though it has harmed most people economically by fostering incredible income disparities. In particular, it has incorporated progressive 1960s discontent with government and, remarkably, dissociated this discontent from its critique of capitalism and corporations.

In a neoliberal society, the market has become an ethics: it has spread everywhere so that all human interactions, from motherhood to education, are discussed as economic "transactions" that can be assessed in individual cost–benefit terms. The market, as Margaret Thatcher argued, "change[s] the soul"[40] by becoming, Foucault argues, the "grid of intelligibility" for everything.[41] This transforms the *homo oeconomicus*—the individual who lies at the base of neoliberalism—from "the [liberal] man of exchange or man the consumer" to "the man of enterprise and consumption."[42] It rests on the "proposition that both parties to an economic transaction benefit from it, *provided the transaction is bi-laterally voluntary and informed*."[43] It focuses on discourses of empowerment in which the worker does not simply own his/her labor, but also possesses his/her own body as a form of "human capital."[44] Since everyone is in control of this form of capital—the body—neoliberalism relies on voluntary, individual actions.[45] Thus, this changed man who has imbibed the market ethic is thus eminently governable, for *homo oeconomicus* is shaped through "rational" and empowering management techniques that make him "self-organized" and "self-controlling."[46]

Relatedly, "user-friendly" computer interfaces have been key to empowering and creating "productive individuals." As Ben Shneiderman, whose work has been key to graphical user interfaces (GUIs), has argued, these interfaces succeed when they move their users from grudging acceptance to feelings of mastery and eagerness.[47] Moreover, this book argues, interfaces—as mediators between the visible and the invisible, as a means of navigation—have been key to creating "informed" individuals who can overcome the chaos of global capitalism by mapping their relation to the totality of the global capitalist system. (Conversely, they enable corporations to track both individuals and totalities, through the data traces produced by our mappings.) The dream is: the resurgence of the *seemingly* sovereign individual, the subject driven to know, driven to map, to zoom in and out, to manipulate, and to act. The dream is: the more that an individual knows, the better decisions he or she can make. Goldman Sachs and other investment companies, for instance, invest millions of dollars on computer programs that can analyze data and execute trades milliseconds faster than their competition. This "informing" is thus intriguingly temporal. New media empowers individuals by informing them of the future, making new media the future. "The future," as William Gibson famously and symptomatically quipped, "is already here. It's just not very evenly distributed."[48] This future—as something that can be bought and sold—is linked intimately to the past, to computers as capable of being the future because, based on past data, they shape and predict it.[49] Computers as future depend on computers as memory machines, on digital data as archives that are always there. This future depends on programmable visions that extrapolate the future—or, more precisely, a future—based on the past. As chapter 1 elaborates, computers, understood as software and hardware machines, have made possible a dream of programmability, a return to a world of Laplaceian determinism in which an all-knowing intelligence can comprehend the future by apprehending the past and present. They have done so through a conflation of words with things that both externalizes knowledge and creates a position from which a subject can try to "hack" the invisible hands and laws that drive the system.

This book, therefore, links computers to governmentality neither at the level of content nor in terms of the many governmental projects that they have enabled, but rather at the level of their architecture and their instrumentality.[50] Computers embody a certain logic of governing or steering through the increasingly complex world around us. By individuating us and also integrating us into a totality, their interfaces offer us a form of mapping, of storing files central to our seemingly sovereign—empowered—subjectivity. By interacting with these interfaces, we are also mapped: data-driven machine learning algorithms process our collective data traces in order to discover underlying patterns (this process reveals that our computers are now more profound programmers than their human counterparts). This logic of programmability, it also argues, is not limited to computer technology; it also stems from and bleeds elsewhere, in particular modern genetics, with its conceptualization of codes and of programs as central to inheritance. Crucially, though, this knowledge is also based on a profound ignorance or ambiguity: our computers execute in unforeseen ways, the future opens to the unexpected. Because of this, any programmed vision will always be inadequate, will always give way to another future. The rest of this book unpacks this temporality and the odd combination of visibility and invisibility these visions enable.

In part I, chapters 1 and 2 focus on how software is invisibly visible. Chapter 1 argues that software emerged as a thing—as an iterable textual program—through an axiomatic process of commercialization and commodification that has made code *logos*: a word conflated with and substituting for *action*. This formulation of instruction as source—source code as fetish—is crucial to understanding the power and thrill of programming, in particular the fantasy of the all-powerful programmer, a subject with magical powers to transform words into things. This separation of code from execution, however, itself a software effect, is also constantly undone, historically and theoretically. Thus, it concludes by analyzing how code as fetish can open up surprising detours and ends.

Chapter 2 analyzes how this invisibly visible (or visibly invisible) logic works at the level of the interface, at the level of "personal computing." It investigates the extent to which this paradoxical combination of rational causality and profound ignorance grounds the computer as an attractive model for the "natural" world. Looking both at the use of metaphor within the early history of human–computer interfaces and at the emergence of the computer as metaphor, it contends that real-time computer interfaces are a powerful response to, and not simply an enabler or consequence of, postmodernism and neoliberalism. Both conceptually and thematically, these interfaces offer a simpler, more reassuring analog of power, one in which the user takes the place of the sovereign "source," code becomes law, and mapping produces the subject.

Chapters 3 and 4 of part II examine the intertwining of computer technology and biology, specifically the emergence of memory and its importance to notions of programmability. Through this focus on the relation between biology and computing technology, part II explores how software, as an axiomatic, came to embody the logic of the "always already there." By exploring the ways in which biology and computer technology have become complementary strands of a double helix, chapters 3 and 4 embed computer technology within the larger epistemological field of programmability, a larger drive for "permanence" that conflates memory with storage and conflates the ephemeral with the enduring, or rather turns the ephemeral into the enduring (the enduring ephemeral) through a process of constant regeneration.

Chapter 3 argues that software was not foreseen, because the drive for software—for an independent program that conflates legislation with execution—did not arise solely from within the field of computation, but also from early Mendelian genetic and eugenics. Through a reading of Erwin Schrödinger's *What Is Life*, it contends that Mendelian genetics and software envision a return to a reductionist, mechanistic understanding of life, in which the human body becomes an archive. This chapter thus complicates the standard narrative within the history of science that the notion of a program was adapted by biologists from computer science, a narrative that rather remarkably treats software as though it always already existed. It also shows how computers, not just in terms of content but also of form, are deeply intertwined with questions of biopower.

The final chapter takes up this intertwining of biology and computer technology, specifically in terms of memory and transmission. Revising the running hypothesis of the first three chapters, chapter 4 shows how digital hardware, which grounds software, is itself axiomatic. Through the reading of early work on neural nets and of John von Neumann's work on automata, it reveals how logical hardware reduces events to words. Analyzing the importance of the analog to conceptualizing the digital, it argues that the digital emerged as a clean, precise logic through an analogy to an analogy. Crucially, it argues that computer memory, as a constantly regenerating and degenerating archive, does not simply erase human agency, but rather makes possible new dreams of human intervention and responsibility.

As this synopsis hopefully makes clear, understanding software as a thing does not mean denigrating software or dismissing it as an ideological construction that covers over the "truth" of hardware. It means engaging its odd materializations and visualizations closely and refusing to reduce software to codes and algorithms—readily readable objects—by grappling with its simultaneous ambiguity and specificity. As Bill Brown has influentially argued, things designate "the concrete yet ambiguous within the everyday," that is, the thing "functions to overcome the loss of other words or as a place holder for some future specifying operation. . . . It designates an amorphous characteristic or a frankly irresolvable enigma. . . . *Things* is a word that tends, especially at its most banal, to index a certain limit or liminality, to hover over the threshold between the nameable and unnameable, the figureable and unfigureable, the identifiable and unidentifiable."[51] Things thus "lie both at hand and somewhere outside the theoretical field, beyond a certain limit, as a recognizable yet illegible remainder or as the entifiable that is unspecifiable."[52] Because things simultaneously name the object and something else, they are both reducible to and irreducible to objects.[53] Whereas we "look *through* objects (to see what they disclose about history, society, nature, or culture—above all, what they disclose about *us*)," we "only catch a glimpse of things."[54] We encounter, but do not entirely comprehend, things.[55] According to Brown:

> A *thing* . . . can hardly function as a window. We begin to confront the thingness of objects when they stop working for us: when the drill breaks, when the car stalls, when the windows get filthy, when their flow within the circuits of production and distribution, consumption and exhibition, has been arrested, however momentarily. The story of objects asserting themselves as things, then, is the story of a changed relation to the human subject and thus the story of how the thing really names less an object than a particular subject-object relation.[56]

Crucially, this effort to rethink, and indeed theorize things, is intimately intertwined with media: Martin Heidegger begins "The Thing" by outlining the shrinking of time and space due to "instant information" (television being the peak of this abolition of every possibility of remoteness); Brown argues, "if the topic of things attained a new urgency in the closing decades of that [twentieth] century, this may have been a response to the digitization of our world—just as, perhaps, the urgency in the 1920s was a response to film."[57]

This book sees this renewed interest in things, things which always seem to be disappearing, not simply as an effect of new media on other "things," but rather as central to the temporality of new media itself. *New media, like the computer technology on which it relies, races simultaneously toward the future and the past, toward the bleeding edge of obsolescence.* Software as thing is inseparable from the externalization of memory, from the dream and nightmare of an all-encompassing archive that constantly regenerates and degenerates, that beckons us forward and disappears before our very eyes.

You

You. Everywhere you turn, it's all about you—and the future. You, the produser. Having turned
off the boob tube, or at least added YouTube, you collaborate, you communicate, you link in,
you download, and you interact. Together, with known, unknown, or perhaps unknowable
others you tweet, you tag, you review, you buy, and you click, building global networks, building community, building databases upon databases of traces. You are the engine behind new
technologies, freely producing content, freely building the future, freely exhausting yourself
and others. Empowered. In the cloud. Telling Facebook and all your "friends" what's on your
mind. Who needs surveillance when you constantly document your life?

But, who or what are you? You are you, and so is everyone else. A shifter, you both
addresses you as an individual and reduces you to a you like everyone else. It is also singular
and plural, thus able to call you and everyone else at the same time. Hey you. Read this.
Tellingly, your home page is no longer that hokey little thing you created after your first HTML
tutorial; it's a mass-produced template, or even worse, someone else's home page—Google's,
Facebook's, the New York Times'. You: you and everyone; you and no one.

1 Invisibly Visible, Visibly Invisible

When enough seemingly insignificant data is analyzed against billions of data elements, the invisible becomes visible.

—Seisint[1]

Computers have fostered both a decline in and frenzy of visual knowledge. Opaque yet transparent, incomprehensible yet logical, they reveal that the less we know the more we show (or are shown). Two phenomena encapsulate this nicely: the proliferation of digital images (new media as "visual culture") and "total information" systems (new media as "transparent").

When digital cameras were introduced to the mass market in the 1990s, many scholars and legal experts predicted the end of photography and film.[2] The reasons they offered were both material and functional: the related losses of celluloid and of indexicality, the evidentiary link between artifact and event. If, as Roland Barthes argues, the photograph certifies that something has been—it is not a "copy" of a past reality, but an "emanation of a *past reality*"[3]—and if, as Mary Ann Doane contends, film as a historical artifact and the filmic moment as historical event are inextricably intertwined,[4] digital images by contrast break the temporal link between record and event. Because a memory card can be constantly rewritten, there is, theoretically, no fixed relationship between captured event and image. Thus, it is not just that digital images are easily manipulated, but also that the moments they refer to cannot be chemically verified. Digital images, in other words, challenge photorealism's conflation of truth and reality: the notion that what is true is what is real and what is real is what is true.

Digital photographs, however, are hardly divorced from either the true or the real, although they relate to them differently than did their celluloid predecessors. Truth is not necessarily coupled to images captured with minimal machinic intervention, but rather to images subject to high-tech manipulation. The so-called *CSI* effect exemplifies this: because of the popular valorization of "forensic" identificatory techniques over deduction, juries are increasingly unwilling to convict based on circumstantial evidence.[5] In addition, although digital photographs were initially treated with suspicion because they were difficult to authenticate, they are now routinely used as evidence both legally and colloquially in part due to their ubiquity: digital images and devices have proliferated wildly. A critical literacy or smartness, verging on paranoia, has also accompanied their use as evidenced by user-driven investigations revealing the darkening of O. J. Simpson's mug shot by *Time Magazine*, the darkening of skies over war-torn Lebanon during the 2006 Isreal-Lebanon conflict by Adnan Hajj, and Dan Rather's unintentional use of forged documents in his investigation of President George W. Bush's war record.

This proliferation, paradoxically, has also fostered a growing belief that computers enable total transparency. Jean Baudrillard in *The Ecstasy of Communication* has argued *"we no longer partake of the drama of alienation, but are in the ecstasy of communication. And this ecstasy is obscene,"* because "in the raw and inexorable light of information," everything is "immediately transparent, visible, exposed."[6] Although extreme, Baudrillard's assessment resonates with public outrage over projects such as the George W. Bush administration's Total Information Awareness Program (TIA), a "systems-level" program developed by the Defense Advanced Research Projects Agency's (DARPA's) Information Awareness Office (IAO) to create a virtual, centralized database, drawing from multiple sources, that would enable the government to capture a person's "information signature." The IAO's motto—*scientia est potentia* (knowledge is power)—and its logo resonated strongly with dystopian science fiction: an eye affixed to the apex of a pyramid, shining a ray of light onto the globe (figure I.1). At all levels, TIA was to enable "topsight": "the ability to 'see the whole thing'—and to plunge in and explore the details."[7] Renamed the Terrorism Information Awareness Program, the funding for this agency was partly revoked by Congress in 2003 in response to citizen complaints, although many of the TIA initiatives, as of 2009, were still funded.

Figure I.1
Information Awareness Office logo

Crucially, this desire to bring together billions of data items was and is not limited to governmental organizations. Google allegedly stores the search terms, linked to IP addresses, of every search on its site; its cameras, designed to produce images for its street view, cruise streets around the world; its "interest-based advertising" monitors user activity in order to refine ads (a technique described by Tim Berners-Lee as similar to allowing someone "to put a television camera in your room, except it will tell them a whole lot more about you than the television camera.")[8] Also, according to the 2009 "KnowPrivacy" report by Joshua Gomez, Travis Pinnick, and Ashkan Soltani of UC Berkeley's iSchool, Google has "a web bug on 92 of the top 100 sites, and on 88% of the total domains reported in the data set of almost 400,000 unique domains."[9] Although Google claims that it does not aggregate these data into one large database, its tracking of consumers through Doubleclick and Google Analytics means that even people who avoid google.com are still tracked by Google. Google—and the Internet— are not the only sites of commercial surveillance. Cable companies use programs like "The Visible World" to target television advertisements to households based on consumption pattern information gathered by firms such as Experian.

This notion of the computer as rendering everything transparent, however, is remarkably at odds with the actual operations of computation, for computers—their hardware, software, and the voltage differences on which they rely—are anything but transparent. When the computer does let us "see" what we cannot normally see, or even when it acts like a transparent medium through video chat, it does not simply relay what is on the other side: it computes. In order to become transparent, the fact that computers always *generate* text and images rather than merely represent or reproduce what exists elsewhere must be forgotten. The current prominence of transparency in product design and in political and scholarly discourse is a compensatory gesture. As our machines increasingly read and write without us, as our machines become more and more unreadable so that seeing no longer guarantees knowing (if it ever did), we the so-called users are offered more to see, more to read. As our machines disappear, getting flatter and flatter, the density and opacity of their computation increases. Every use is also an act of faith: we believe these images and systems render us transparent not for technological, but rather for metaphorical, or more strongly ideological, reasons.

As stated earlier, this paradox is not accidental to computing's appeal, but rather grounds the computer as a useful and provocative, indeed magical, model. Its combination of what can be seen and not seen, can be known and not known—its separation of interface from algorithm; software from hardware—makes it a powerful metaphor for everything we believe is invisible yet generates visible effects, from genetics to the invisible hand of the market; from ideology to culture. Joseph Weizenbaum has argued that computers have become metaphors for all "effective procedures," that is, for anything that can be solved in a prescribed number of steps, such as gene expression and clerical work.[10] Weizenbaum also notes that the computer as metaphor is itself based on "only the vaguest understanding of a difficult and complex scientific concept. . . . The public vaguely understands—but is nonetheless firmly convinced—that any effective procedure can, in principle, be carried out by a computer."[11] Even a computer programmer, Weizenbaum notes, cannot "know the path of decision making within his own program, let alone what intermediate or final results it will produce."[12] But critiques—even those as insightful as Joseph Weizenbaum's—that condemn the computer as a poor model because of its contradictory reductionism and incomprehensibility miss the point. Revealing the illogical intertwining of computers we cannot understand with understanding will not dispel the power of the computer as metaphor because this intertwining grounds its appeal. The linking of rationality with mysticism, knowability with what is unknown, makes it a powerful fetish that offers its programmers and users alike a sense of empowerment, of sovereign subjectivity, that covers over—barely—a sense of profound ignorance.

The following two chapters address this causal pleasure through software, or, to be more precise, the curious separation of software from hardware. Software perpetuates certain notions of seeing as knowing, of reading and readability, which were supposed to have faded with the waning of indexicality, by producing WYSIWG (What You See Is What You Get) interfaces that mimic both ideology *and* ideology critique, the process of covering and uncovering.[13] As I explain in more detail in chapter 2, it offers us a way to cognitively map our increasingly complex world, or at least to understand, often pleasurably, our relation to its complexity. Software, through programming languages that stem from a gendered system of command and control, creates an invisible system of visibility, a system of causal pleasure. This system renders our machine's normal processes demonic and makes our computer truly a medium: something in between, mystical, channeling, and not entirely trustworthy. It becomes a conduit that also amplifies and selects what is at once real and unreal, true and untrue, visible and invisible.

1 On Sourcery and Source Codes

The spirit speaks! I see how it must read,
And boldly write: "In the beginning was the Deed!"
—Johann Wolfgang Goethe[1]

Software emerged as a thing—as an iterable textual program—through a process of commercialization and commodification that has made code *logos*: code as source, code as true representation of action, indeed, code as conflated with, and substituting for, action.[2] Now, in the beginning, is the word, the instruction. Software as logos turns *program* into a noun—it turns process in time into process in (text) space. In other words, Manfred Broy's software "pioneers," by making software easier to visualize, not only sought to make the implicit explicit, they also created a system in which the intangible and implicit drives the explicit. They thus obfuscated the machine and the process of execution, making software the end all and be all of computation and putting in place a powerful logic of sourcery that makes source code—which tellingly was first called pseudocode—a fetish.[3]

This chapter investigates the implications of code as logos and the ways in which this simultaneous conflation and separation of instruction from execution, itself a software effect, is constantly constructed and undone, historically and theoretically. This separation is crucial to understanding the power and thrill of programming, in particular the nostalgic fantasy of an all-powerful programmer, a sovereign neoliberal subject who magically transforms words into things. It is also key to addressing the nagging doubts and frustrations experienced by programmers: the sense that we are slaves, rather than masters, clerks rather than managers—that, because "code is law," the code, rather than the programmer, rules. These anxieties have paradoxically led to the romanticization and recuperation of early female operators of the 1946 Electronic Numerical Integrator and Computer (ENIAC) as the first programmers, for they, unlike us, had intimate contact with and knowledge of the machine. They did not even need code: they engaged in what is now called "direct programming," wiring connections and setting values. Back then, however, the "master programmer" was part of the machine (it controlled the sequence of calculation); computers, in contrast, were human. Rather than making programmers and users either masters or slaves, code as logos establishes a perpetual oscillation between the two positions: every move to empower also estranges.

This chapter, however, does not call for a return to direct programming or hardware algorithms, which, as I argue in chapter 4, also embody logos. It also does not endorse such a call because the desire for a "return" to a simpler map of power drives source code as logos. The point is not to break free from this sourcery, but rather to play with the ways in which logos also invokes "spellbinding powers of enchantment, mesmerizing fascination, and alchemical transformation."[4] The point is to make our computers more productively spectral by exploiting the unexpected possibilities of source code as fetish. As a fetish, source code produces surprisingly "deviant" pleasures that do not end where they should. Framed as a re-source, it can help us think through the machinic and human rituals that help us imagine our technologies and their executions. The point is also to understand how the surprising emergence of code as logos shifts early and still-lingering debates in new media studies over electronic writing's relation to poststructuralism, debates that the move to software studies has to some extent sought to foreclose.[5] Rather than seeing technology as simply fulfilling or killing theory, this chapter outlines how the alleged "convergence" between theory and technology challenges what we thought we knew about logos. Relatedly, engaging source code as fetish does not mean condemning software as immaterial; rather, it means realizing the extent to which software, as an "immaterial" relation become thing, is linked to changes in the nature of subject-object relations more generally. Software as thing can help us link together minute machinations and larger flows of power, but only if we respect its ability to surprise and to move.

Source Code as Logos

To exaggerate slightly, software has recently been posited as the essence of new media and knowing software a form of enlightenment. Lev Manovich, in his groundbreaking *The Language of New Media*, for instance, asserts: "New media may look like media, but this is only the surface. . . . To understand the logic of new media, we need to turn to computer science. It is there that we may expect to find the new terms, categories, and operations that characterize media that become programmable. *From media studies, we move to something that can be called 'software studies'—from media theory to software theory.*"[6] This turn to software—to the logic of what lies beneath—has offered a solid ground to new media studies, allowing it, as Manovich argues, to engage presently existing technologies and to banish so-called "vapor theory"—theory that fails to distinguish between demo and product, fiction and reality—to the margins.[7]

This call to banish vapor theory, made by Geert Lovink and Alexander Galloway among others, has been crucial to the rigorous study of new media, but this rush away from what is vapory—undefined, set in motion—is also troubling because vaporiness is not accidental but rather essential to new media and, more broadly, to software. Indeed, one of this book's central arguments is that a rigorous engagement with software makes new media studies more, rather than less, vapory. Software, after all, is ephemeral, information ghostly, and new media projects that have never, or barely, materialized are among the most valorized and cited.[8] (Also, if you take the technical definition of information seriously, information increases with vapor, with entropy). This turn to computer science also threatens to reify knowing software as truth, an experience that is arguably impossible: we all know some software, some programming languages, but does anyone really "know" software? What could this knowing even mean? Regardless, from myths of all-powerful hackers who "speak the language of computers as one does a mother tongue"[9] or who produce abstractions that release the virtual[10] to perhaps more mundane claims made about the radicality of open source, knowing (or using the right) software has been made analogous to man's release from his self-incurred tutelage.[11] As advocates of free and open source software make clear, this critique aims at political, as well as epistemological, emancipation. As a form of enlightenment, it is a stance of how not to be governed like that, an assertion of an essential freedom that can only be curtailed at great cost.[12]

Knowing software, however, does not simply enable us to fight domination or rescue software from "evil-doers" such as Microsoft. Software, free or not, is embedded and participates in structures of knowledge-power. For instance, using free software does not mean escaping from power, but rather engaging it differently, for free and open source software profoundly privatizes the public domain: GNU copyleft—which allows one to use, modify, and redistribute source code and derived programs, but only if the original distribution terms are maintained—seeks to fight copyright by spreading licences everywhere.[13] More subtly, the free software movement, by linking freedom and freely accessible source code, amplifies the power of source code both politically and technically. It erases the vicissitudes of execution and the institutional and technical structures needed to ensure the coincidence of source code and its execution. This amplification of the power of source code also dominates critical analyses of code, and the valorization of software as a "driving layer" conceptually constructs software as neatly layered.

Programmers, computer scientists, and critical theorists have reduced software to a recipe, a set of instructions, substituting space/text for time/process. The current common-sense definition of *software* as a "set of instructions that direct a computer to do a specific task" and the OED definition of software as "the programs and procedures required to enable a computer to perform a specific task, as opposed to the physical components of the system" both posit software as cause, as what drives computation. Similarly, Alexander Galloway argues, "code draws a line between what is material and what is active, in essence saying that writing (hardware) cannot *do* anything, but must be transformed into code (software) to be effective. . . . Code is a language, but a very special kind of language. *Code is the only language that is executable* . . . code is the first language that actually does what it says."[14] This view of software as "actually doing what it *says*" (emphasis added) both separates instruction from, and makes software substitute for, execution. It assumes no difference between source code and execution, between instruction and result. That is, Galloway takes the principles of executable layers (application on top of operating system, etc.) and grafts it onto the system of compilation or translation, in which higher-level languages are transformed into executable codes that are then executed line by line. By doing what it "says," code is surprisingly logos. Like the King's speech in Plato's *Phaedrus*, it does not pronounce knowledge or demonstrate it—it transparently pronounces itself.[15] The hidden signified—meaning—shines through and transforms itself into action. Like Faust's translation of logos as "deed," code is action, so that "in the beginning was the Word, and the Word was with God, and the Word was God."[16]

Not surprisingly, many scholars critically studying code have theorized code as performative. Drawing in part from Galloway, N. Katherine Hayles in *My Mother Was a Computer: Digital Subjects and Literary Texts* distinguishes between the linguistic performative and the machinic performative, arguing:

> Code that runs on a machine is performative in a much stronger sense than that attributed to language. When language is said to be performative, the kinds of actions it "performs" happen in the minds of humans, as when someone says "I declare this legislative session open" or "I pronounce you husband and wife." Granted, these changes in minds can and do reach in behavioral effects, but the performative force of language is nonetheless tied to the external changes through complex chains of mediation. By contrast, code running in a digital computer causes changes in machine behavior and, through networked ports and other interfaces, may initiate other changes, all implemented through transmission and execution of code.[17]

The independence of machine action—this autonomy, or automatic executability of code—is, according to Galloway, its material essence: "The material substrate of code, which must always exist as an amalgam of electrical signals and logical operations in silicon, however large or small, demonstrates that code exists first and foremost as commands issued to a machine. Code essentially has no other reason for being than instructing some machine in how to act. One cannot say the same for the natural languages."[18] Galloway thus concludes in "Language Wants to Be Overlooked: On Software and Ideology," "to see code as subjectively performative or enunciative is to anthropomorphize it, to project it onto the rubric of psychology, rather than to understand it through its own logic of 'calculation' or 'command.'"[19]

To what extent, however, can source code be understood outside of anthropomorphization? Does understanding voltages stored in memory as commands/code not already anthropomorphize the machine? The title of Galloway's article, "Language *Wants* to Be Overlooked" (emphasis mine), inadvertently reveals the inevitability of this anthropomorphization. How can code/language want—or most revealingly *say*—anything? How exactly does code "cause" changes in machine behavior? What mediations are necessary for this insightful yet limiting notion of code as inherently executable, as conflating meaning and action?

Crafty Sources

To make the argument that code is automatically executable, the process of execution itself not only must be erased, but source code must also be conflated with its executable version. This is possible, Galloway argues, because the two "layers" of code can be reduced to each other: "uncompiled source code is *logically* equivalent to that same code compiled into assembly language and/or linked into machine code. For example, it is absurd to claim that a certain value expressed as a hexadecimal (base 16) number is more or less fundamental than that same value expressed as binary (base 2) number. They are simply two expressions of the same value."[20] He later elaborates on this point by drawing an analogy between quadratic equations and software layers:

> One should never understand this "higher" symbolic machine as anything empirically different from the "lower" symbolic interactions of voltages through logic gates. They are complex aggregates yes, but it is foolish to think that writing an "if/then" control structure in eight lines of assembly code is any more or less machinic than doing it in one line of C, just as the same quadratic equation may swell with any number of multipliers and still remain balanced. The relationship between the two is *technical*.[21]

According to Galloway's quadratic equation analogy, the difference between a compact line of higher-level programming code and eight lines written in assembler equals the difference between two equations, in which one contains coefficients that are multiples of the other. The solution to both equations is the same: one equation is the same as the other.

This reduction, however, does not capture the difference between the various instantiations of code, let alone the empirical difference between the higher symbolic machine and the lower interactions of voltages (the question here is: where does one make the empirical observation?). To state the obvious, one cannot run source code: it must be compiled or interpreted. This compilation or interpretation—this making executable of code—is not a trivial action; the compilation of code is not the same as translating a decimal number into a binary one. Rather, it involves instruction explosion and the translation of symbolic into real addresses. Consider, for example, the instructions needed for adding two numbers in PowerPC assembly language, which is one level higher than machine language:

```
li    r3,1        *load the number 1 into register 3

li    r4,2        *load the number 2 into register 4

add   r5,r4,r3    *add r3 to r4 and store the result in r5

stw   r5,sum(rtoc) *store the contents of r5 (i.e., 3) into the memory location

                  *called "sum" (where sum is defined elsewhere)

blr               *end of this snippet of code[22]
```

This explosion is not equivalent to multiplying both sides of a quadratic equation by the same coefficient or to the difference between E and 15. It is, instead, a breakdown of the steps needed to perform a simple arithmetic calculation; it focuses on the movement of data within the machine. The relationship between executable and higher-level code is not that of mathematical identity but rather logical equivalence, which can involve a leap of faith. This is clearest in the use of numerical methods to turn integration—a function performed fluidly in analog computers—into a series of simpler, repetitive arithmetical steps.

This translation from source code to executable is arguably as involved as the execution of any command, and it depends on the action (human or otherwise) of compiling/interpreting and executing. Also, some programs may be executable, but not all compiled code within that program is executed; rather, lines are read in as necessary. Software is "layered" in other words, not only because source is different from object, but also because object code is embedded within an operating system.

So, to spin Galloway's argument differently, a technical relation is far more complex than a numerical one. Rhetoric was considered a *technê* in antiquity. Drawing on this Paul Ricoeur explains, "*technê* is something more refined than a routine or an empirical practice and in spite of its focus on production, it contains a speculative element."[23] A technical relation engages art or craft. A technical person is one "skilled in or practically conversant with some particular art or subject."[24] Code does not always or automatically do what it says, but it does so in a crafty, speculative manner in which meaning and action are both created. It carries with it the possibility of deviousness: our belief that compilers simply expand higher-level commands—rather than alter or insert other behaviors—is simply that, a belief, one of the many that sustain computing as such. This belief glosses over the fact that *source code only becomes a source after the fact*. Execution, and a whole series of executions, belatedly makes some piece of code a source, which is again why source code, among other things, was initially called pseudocode.

Source code is more accurately a *re-source*, rather than a source. Source code becomes the source of an action only after it—or more precisely its executable substitute—expands to include software libraries, after its executable version merges with code burned into silicon chips; and after all these signals are carefully monitored, timed, and rectified. Source code becomes a source only through its destruction, through its simultaneous nonpresence and presence.[25] (Thus, to return to the historical difficulties of analyzing software outlined by Mahoney, every software run is to some extent a reconstruction.) Source code as *technê*, as a generalized writing, is spectral. It is neither dead repetition nor living speech; nor is it a machine that erases the difference between the two. It, rather, puts in place a "relation between life and death, between present and representation, between two apparatuses."[26] As I elaborate throughout this book, information—through its capture in memory—is undead.

Source Code, after the Fact

Early on, the difficulties of code as source were obvious. Herman H. Goldstine and John von Neumann emphasized the dynamic nature of code in their "Planning and Coding of Problems for an Electronic Computing Instrument." In it, they argued that coding, despite the name, is not simply the static translation of "a meaningful text (the instructions that govern solving the problem under consideration) from one language (the language of mathematics, in which the planner will have conceived the problem, or rather the numerical procedure by which he has decided to solve the problem) into another language (that of our code)."[27] Because code does not unfold linearly, because its value depends on intermediate results, and because code can be modified as it is run (self-modifying code), "it will not be possible in general to foresee in advance and completely the actual course of C [the sequence of codes]." Therefore, "coding is . . . the technique of providing a dynamic background to control the automatic evolution of a meaning."[28] Code as "dead repetition," in other words, has always been regenerative and interactive; every iteration alters its meaning. Even given the limits to iterability that Hayles has presciently outlined in *My Mother Was a Computer*—limits due to software as axiomatic—coding still means producing a mark, a writing, open to alteration/iteration rather than an airtight anchor.[29]

Much disciplinary effort has been required to make source code readable as the source. Structured programming, which I examine in more detail later, sought to rein in "goto crazy" programmers and self-modifying code. A response to the much-discussed "software crisis" of the late 1960s, its goal was to move programming from a craft to a standardized industrial practice by creating disciplined programmers who dealt with abstractions rather than numerical processes.[30]

Making code the source also entails reducing hardware to memory and thus erasing the existence and possibility of hardware algorithms. Code is also not always the source because hardware does not need software to "do something." One can build algorithms using hardware. Figure 1.1, for instance, is the logical statement: if notB and notA, do CMD1 (state P); if notB and notA and notZ OR B and A (state Q) then command 2.

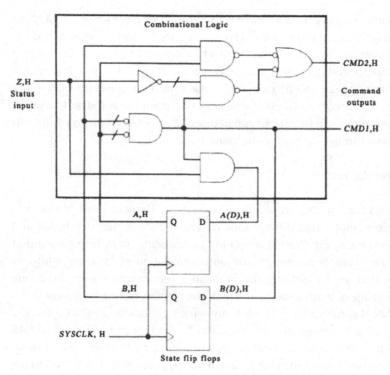

Figure 1.1
Logic diagram for a hardware algorithm

To be clear, I am not valorizing hardware over software, as though hardware naturally escapes this drive to make space signify time. Crucially, this schematic is itself an abstraction. Logic gates can only operate "logically"—as logos—if they are carefully timed. As Philip Agre has emphasized, the digital abstraction erases the fact that gates have "directionality in both space (listening to its inputs, driving its outputs) and in time (always moving toward a logically consistent relation between these inputs and outputs)."[31] When a value suddenly changes, there is a brief period in which a gate will give a false value. In addition, because signals propagate in time over space, they produce a magnetic field that can corrupt other nearby signals (known as *crosstalk*). This schematic erases all these various time- and distance-based effects by rendering space blank, empty, and banal. Thus hardware schematics, rather than escaping from the logic of sourcery, are also embedded within this structure. Indeed, as chapter 4 elaborates, John von Neumann, the generally acknowledged architect of the stored-memory digital computer, drew from Warren McCulloch and Walter Pitts's conflation of neuronal activity with its inscription in order to conceptualize modern computers. It is perhaps appropriate then that von Neumann, who died from a cancer stemming from his work at Los Alamos, spent the last days of his life reciting from memory *Faust Part 1*.[32] At the source of stored program computing lies the Faustian erasure of word for action.

The notion of source code as source coincides with the introduction of alphanumeric languages. With them, human-written, nonexecutable code becomes source code and the compiled code, the object code. Source code thus is arguably symptomatic of human language's tendency to attribute a sovereign source to an action, a subject to a verb.[33] By converting action into language, source code emerges. Thus, Galloway's statement, "To see code as subjectively performative or enunciative is to anthropomorphize it, to project it onto the rubric of psychology, rather than to understand it through its own logic of 'calculation' or 'command,'" overlooks the fact that to use higher-level alphanumeric languages is already to anthropomorphize the machine. It is to embed computers in "logic" and to reduce all machinic actions to the commands that supposedly drive them. In other words, the fact that "code is law"—something legal scholar Lawrence Lessig emphasizes—is hardly profound.[34] After all, code is, according to the OED, "a systematic collection or digest of the laws of a country, or of those relating to a particular subject." What is surprising is the fact that software is code; that code is—has been made to be—executable, and this executability makes code not law, but rather every lawyer's dream of what law should be: automatically enabling and disabling certain actions, functioning at the level of everyday practice.[35]

Code is executable because it embodies the power of the executive, the power of enforcement that has traditionally—even within classic neoliberal logic—been the provenance of government.[36] Whereas neoliberal economist and theorist Milton Friedman must concede the necessity of government because of the difference between "the day-to-day activities of people [and] the general customary and legal framework within which these take place," code as self-enforcing law "privatizes" this function, further reducing the need for government to enforce the rules by which we play.[37] In other words, if as Foucault argues neoliberalism expands judicial interventions by reducing laws to "the rules for a game in which each remains master regarding himself and his part," then "code is law" reins in this expansion by moving enforcement from police and judicial functions to software functions.[38] "Code is law," in other words, automatically brings together disciplinary and sovereign power through the production of self-enforcing rules that, as von Neumann argues, "govern" a situation.

"Code is law" makes clear the desire for sovereign power driving both source code and performative utterances more generally. David Golumbia—looking more generally at widespread beliefs about computers—has insightfully claimed: "The computer encourages a Hobbesian conception of this political relation: one is either the person who makes and gives orders (the sovereign), or one follows orders."[39]

This conception, which crucially is also constantly undone by modern computation's twinning of empowerment with ignorance, depends, I argue, on this conflation of code with the performative. As Judith Butler has argued in *Excitable Speech*, Austinian understandings of performative utterances as simply doing what they say posit the speaker as "the judge or some other representative of the law."[40] It resuscitates fantasies of sovereign—that is *executive* (hence executable)—structures of power: it is "a wish to return to a simpler and more reassuring map of power, one in which the assumption of sovereignty remains secure."[41] This wish for a simpler map of power—indeed power as mappable—drives not only code as automatically executable, but also, as the next chapter contends, interfaces more generally. This wish is central to computers as machines that enable users/programmers to navigate neoliberal complexity.

Against this nostalgia, Butler, following Jacques Derrida, argues that iterability lies behind the effectiveness of performative utterances. For Butler, iterability is the process by which *"the subject who 'cites' the performative is temporarily produced as the belated and fictive origin of the performative itself."*[42] The programmer/user, in other words, is produced through the act of programming. Moreover, the effectiveness of performative utterances, Butler also emphasizes, is intimately tied to the community one joins and to the rituals involved—to the history of that utterance. Code as law—as a judicial process—is, in other words, far more complex than code as logos. Similarly, as Weizenbaum has argued, code understood as a judicial process undermines the control of the programmer:

> A large program is, to use an analogy of which Minsky is also fond, an intricately connected network of courts of law, that is, of subroutines, to which evidence is transmitted by other subroutines. These courts weigh (evaluate) the data given to them and then transmit their judgments to still other courts. The verdicts rendered by these courts may, indeed, often do, involve decisions about what court has "jurisdiction" over the intermediate results then being manipulated. The programmer thus cannot even know the path of decision-making within his own program, let alone what intermediate or final results it will produce. Program formulation is thus rather more like the creation of a bureaucracy than like the construction of a machine of the kind Lord Kelvin may have understood.[43]

Code as a judicial process is code as *thing*: the Latin term for thing, *res*, survives in legal discourse (and, as I explain later, literary theory). The term *res*, as Heidegger notes, designates a "gathering," any thing or relation that concerns man.[44] The relations that Weizenbaum discusses, these bureaucracies within the machine, as the rest of this chapter argues, mirror the bureaucracies and hierarchies that historically made computing possible. Importantly, this description of computers as following a set of rules that programmers must follow—Weizenbaum's insistence on the programmer's ignorance—does not undermine the resonances between neoliberalism and computation; if anything, it makes these resonances more clear. It also clarifies the desire driving code as logos as a solution to neoliberal chaos. Foucault, emphasizing the rhetoric of the economy as a "game" in neoliberal writings, has argued, "both for the state and for individuals, the economy must be a game: a set of regulated activities . . . in which the rules are not decisions which someone takes for others. It is a set of rules which determine the way in which each must play a game whose outcome is not known by anyone."[45] Although small-s sovereigns proliferate through neoliberalism's empowered yet endangered subjects, it still fundamentally denies the position of the Sovereign who knows—a position that we nonetheless nostalgically desire . . . for ourselves.

Yes, Sir!

This conflation of instruction with result stems in part from software's and computing's gendered, military history: in the military there is supposed to be no difference between a command given and a command completed—especially to a computer that is a "girl." For computers, during World War II, were in fact young women with some background in mathematics. Not only were women available for work during that era, they also were considered to be better, more conscientious computers, presumably because they were better at repetitious, clerical tasks. They were also undifferentiated: they were all unnamed "computers," regardless of their mathematical training.[46] These computers produced ballistics tables for new weapons, tables designed to control servicemen's battlefield actions. Rather than aiming and shooting, servicemen were to set their guns to the proper values (not surprisingly, these tables and gun governors were often ignored or ditched by servicemen).[47]

The women who became the "ENIAC girls" (later the more politically correct "women of the ENIAC")—Kathleen/Kay McNulty (Mauchly Antonelli), Jean Jennings (Bartik), Frances Snyder (Holberton), Marlyn Wescoff (Meltzer), Frances Bilas (Spence), and Ruth Lichterman (Teitelbaum) (married names in parentheses)—were computers who volunteered to work on a secret project (when they learned they would be operating a machine, they had to be reassured that they had not been demoted). Programmers were former computers because they were best suited to prepare their successors: they thought and acted like computers. One could say that programming became programming and software became software when the command structure shifted from commanding a "girl" to commanding a machine. Kay Mauchly Antonelli described the "evolution" of computing as moving from female computers using Marchant machines to fill in fourteen-column sheets (which took forty hours to complete the job), to using differential analyzers (fifteen minutes to do the job), to using the ENIAC (seconds).[48]

Software languages draw from a series of imperatives that stem from World War II command and control structures. The automation of command and control, which

Paul Edwards has identified as a perversion of military traditions of "personal leader-ship, decentralized battlefield command, and experience-based authority,"[49] arguably started with World War II mechanical computation. Consider, for instance, the rela-tionship between the volunteer members of the Women's Royal Naval Service (called Wrens), and their commanding officers at Bletchley Park. The Wrens also (perhaps ironically) called *slaves* by the mathematician and "founding" computer scientist Alan Turing (a term now embedded within computer systems), were clerks responsible for the mechanical operation of the cryptanalysis machines (the Bombe and then the Colossus), although at least one of the clerks, Joan Clarke (Turing's former fiancé), became an analyst. Revealingly, I. J. Good, a male analyst, describes the Colossus as enabling a man–machine synergy duplicated by modern machines only in the late 1970s: "the analyst would sit at the typewriter output and call out instructions to a Wren to make changes in the programs. Some of the other uses were eventually reduced to decision trees and were handed over to the machine operators (Wrens)."[50] This man–machine synergy, or interactive real-time (rather than batch) processing, treated Wrens and machines indistinguishably, while simultaneously relying on the Wrens' ability to respond to the mathematician's orders. This "interactive" system also seems evident in the ENIAC's operation: in figure 1.2, a male analyst issues commands to a female operator.

The story of the initial meeting between Grace Murray Hopper (one of the first and most important programmer-mathematicians) and Howard Aiken would also seem to buttress this narrative. Hopper, with a PhD in mathematics from Yale, and a former mathematics professor at Vassar, was assigned by the U.S. Navy to program the Mark 1, an electromechanical digital computer that made a sound like a roomful of knitting needles. According to Hopper, Aiken showed her "a large object with three stripes . . . waved his hand and said: 'That's a computing machine.' I said, 'Yes, Sir.' What else could I say? He said he would like to have me compute the coefficients of the arc tangent series, for Thursday. Again, what could I say? 'Yes, Sir.' I didn't know what on earth was happening, but that was my meeting with Howard Hathaway Aiken."[51] Computation depends on "Yes, Sir" in response to short declarative sentences and imperatives that are in essence commands. Contrary to Neal Stephenson, in the beginning—marking the possibility of a beginning—was the command rather than the command line.[52] The command line is a mere operating system (OS) simulation. Com-mands have enabled the slippage between programming and action that makes soft-ware such a compelling yet logically "trivial" communications system.[53] Commands lie at the core of the cybernetic conflation of human with machine.[54] I. J. Good's and Hopper's recollections also reveal the routinization at the core of programming: the analyst's position at Bletchley Park was soon replaced by decision trees acted on by the Wrens. Hopper, self-identified as a mathematician (not programmer), became an advocate of automatic programming. Thus routinization or automation lies at the

Figure 1.2
ENIAC programmers, late 1940s. U.S. military photo, Redstone Arsenal Archives, Huntsville, Alabama.

core of a profession that likes to believe it has successfully automated every profession but its own.[55]

This narrative of the interchangeability of women and software, however, is not entirely true: the perspective of the master, as Hegel famously noted, is skewed. (Tellingly, Mephistopheles offers to be Faust's servant.)[56] The master depends on the slave entirely, and it is the slave's actions that make possible another existence. Execution is never simple. Hopper's "Yes, Sir" actually did follow in the military command tradition. It was an acceptance of responsibility; she was not told how to calculate the trajectory. Also, the "women of the ENIAC," although an afterthought, played an important role in converting the ENIAC into a stored-program computer and in determining the trade-off between storing values and instructions: they did not simply operate the machine, they helped shape it and make it functional.[57] Users of the ENIAC usually were divided into pairs: one who knew the problem and one who knew the machine "so the limitations of the machine could be fitted to the problem and the problem could be changed to fit the limitations."[58] Programming the ENIAC—that is, wiring the components together in order to solve a problem—was difficult, especially since there were no manuals or exact precedents.[59] To solve a problem, such as how to determine ballistics trajectories for new weapons, ENIAC "programmers" had first to break down the problem logically into a series of small yes/no decisions; "the amount of work that had to be done before you could ever get to a machine that was really doing any thinking," Bartik relates, was staggering and annoying.[60] The unreliability of the hardware and the fact that engineers and custodians would unexpectedly change the switches and program cables compounded the difficulty.[61]

These women, Holberton in particular, developed an intimate relation with the "master programmer," the ENIAC's control device. Although Antonelli first figured out how to repeat sections of the program, using the master programmer, Holberton, who described herself as a logician, specialized in controlling its operation.[62] As Bartik explains:

> We found it very easy to learn that you do this step, step one, then you do step two, step three, but I think the thing that was the hardest for us to learn was transfer of control which the ENIAC did have through the master programmer, so that you would be able to repeat pieces of program. So, the techniques for dividing your program into subroutines that could be repeated and things of this kind was the hardest for us to understand. I certainly know it was for me.[63]

Because logic diagrams did not then exist, Holberton developed a four-color pencil system to visualize the workings of the master programmer.[64] This drive to visualize also extended to the machine as a whole. To track the calculation, holes were drilled in the panels over the accumulators so that "when you were doing calculations these lights were flashing as the numbers built up and as you transferred numbers and things of this kind. So you had the feeling of excitement."[65] These lights not only were useful in tracking the machine, they also were invaluable for the demonstration. Even though the calculation for the demonstration was itself buggy, the flashing lights, the cards being read and written, gave the press a (to them) incomprehensible visual display of the enormity and speed of the calculation being done. In what would become a classic programming scenario, the problem was "debugged" the day after the demonstration. According to Holberton:

> I think the next morning, I woke up and in the middle of the night thinking what that error was. I came in, made a special trip on the early train that morning to look at a certain wire, and you know, it's the same kind of programming error that people make today. It's the, the decision on the terminal end of a do loop, speaking Fortran language, had the wrong value. Forgetting that zero was also one setting and the setting of the switch was one off. And I'll never forget that because there it was my first do loop error. But it went on that way and I remember telling Marlyn, I said, "If anybody asks why it's printing out that way, say it's supposed to be that way." [Laughter][66]

Programming enables a certain duplicity, as well as the possibility of endless actions that animate the machine. Holberton, described by Hopper as the best programmer she had known, would also go on to develop an influential SORT algorithm for the UNIVAC 1 (the Universal Automatic Computer 1, a commercial offshoot of the ENIAC).[67] Indeed, many of these women were hired by the Eckert–Mauchly company to become the first programmers of the UNIVAC, and were transferred to Aberdeen to train more ENIAC programmers.

Drawing from the historical importance of women and the theoretical resonances between the feminine and computing (parallels between programming and what Freud called the quintessentially feminine invention of weaving, between female sexuality as mimicry and Turing's vision of computers as universal machines/mimics) Sadie Plant has argued that computing is essentially feminine. Both software and feminine sexuality reveal the power that something that cannot be seen can have.[68] Women, Plant argues, "have not merely had a minor part to play in the emergence of digital machines. . . . Theirs is not a subsidiary role which needs to be rescued for posterity, a small supplement whose inclusion would set the existing records straight. . . . Hardware, software, wetware—before their beginnings and beyond their ends, women have been the simulators, assemblers, and programmers of the digital machines."[69] Because of this and women's early (forced) adaptation to "flexible" work conditions, Plant argues, women are best prepared to face our digital, networked future: "sperm count," she writes, "falls as the replicants stir and the meat learns how to learn for itself. Cybernetics is feminisation."[70] Responding to Plant's statement, Alexander Galloway has argued, "the universality of [computer] protocol can give feminism something that it never had at its disposal, the obliteration of the masculine from beginning to end."[71] Protocol, Galloway asserts, is inherently antipatriarchy. What, however, is the relationship between feminization and feminism, between so-called feminine modes of control and feminism? What happens if you take seriously Grace Murray Hopper's claims that the term *software* stemmed from her description of compilers as "layettes" for computers and the claim of J. Chuan Chu, one of the hardware engineers for the ENIAC, that software is the "daughter" of Frankenstein (hardware being the son)?[72]

To address these questions, we need to move beyond recognizing these women as programmers and the resonances between computers and the feminine. Such recognition alone establishes a powerful sourcery, in which programming is celebrated at the exact moment that programmers become incapable of "understanding"—of seeing through—the machine. The move to reclaim the ENIAC women as the first programmers in the mid- to late-1990s occurred when their work as operators—and the visual, intimate knowledge of machine operations this entailed—had become entirely incorporated into the machine and when women "coders" were almost definitively pushed out of the workplace. It is love at last (and first) sight, not just for these women but also for these interfaces, which really were transparent holes, in which inside and

outside coincided. Also, reclaiming these women as the first programmers and as
feminist figures glosses over the hierarchies within programming—among operators,
coders, and analysts—that defined the emergence of programming as a profession and
as an academic discipline.[73] To put Hopper and the "ENIAC girls" together is to erase
the difference between Hopper, a singular hero who always defined herself as a mathematician, and nameless disappearing computer operators. It is also to deny personal
history: Hopper, a social conservative from a privileged background, stated many times
that she was not a feminist, and Hopper's stances could be perceived as antifeminist
(while the highest-ranking female officer in the Navy, she argued that women were
incapable of serving in combat duty).[74] Not accidentally, Hopper's dream, her drive
for automatic computing, was to put the programmer inside the computer and thus
to rehumanize the mathematician: pseudocode was to free the mathematician and
her brain from the shackles of programming.[75]

Bureaucracies within the Machine

TROPP: We talked about Von Neumann and I would like to talk about how you saw people like
John Mauchly and the role that they played, and Goldstine and Burks and others that you came
in contact with [including] Clippinger, and Frankel, and how, how they looked from your vantage
point?
HOLBERTON: Well, we were lowly programmers, so I looked up to all these gentlemen.
TROPP: [Laughter][76]

The conflation of instruction with action, which makes computers understood as
software and hardware machines such a compelling model of neoliberal governmentality and which resuscitates dreams of sovereign power, depends on incorporating
historical programming hierarchies within the machine.

   Programming, even at what has belatedly been recognized as its origin, was a hierarchical affair. Herman H. Goldstine and John von Neumann, in "Planning and
Coding of Problems for an Electronic Computing Instrument," separated the task of
planning (dealing with the dynamic nature of code through extensive flow charting)
from that of coding (the microproduction of the actual instructions). Regarding
dynamic or macroscopic aspects, they argued, "every mathematician, or every mod-
erately mathematically trained person should be able to do this in a routine manner,
if he has familiarized himself with the main examples that follow in this report, or if
he has had some equivalent training in this method." Regarding the static or microscopic work, they asserted, "we feel certain that a moderate amount of experience with
this stage of coding suffices to remove from it all difficulties, and to make it a perfectly
routine operation."[77] The dropping of the pronoun *he* was not accidental: as Nathan
Ensmenger and William Aspray note, the dynamic analysis was to be performed by
"the 'planner,' who was typically the scientific user and overwhelmingly often was

male; the sixth task was to be carried out by 'coders'—almost always female."[78] Although this separation between operators, coders, and planners was not immediately accepted everywhere—the small Whirlwind group viewed itself more as a "model shop" in which coding, programming, and operations were mixed together—this hierarchical separation between what Philip Kraft calls the "head and the hand" became dominant as programming became a mass, commercial enterprise.[79]

SAGE (the Semi-Automatic Ground Environment) air defense system, widely considered the first large software project, was programmed by the Systems Development Corporation (SDC), an offshoot of the RAND Corporation. SDC had expanded from a few programmers to more than eight hundred by the late 1950s, making it by far the largest employer of programmers. Because its programmers went on to form the industry (it was dubbed the "university of programmers"), SAGE had a wide impact on the field's development. SAGE, however, not only taught people how to code but also inculcated a strict division of programming in which senior programmers (later systems analysts), who developed program specifications, were separated from programmers, who worked on coding specifications; they in turn were separated from the coders who turned coding specifications into documented machine code.[80] This separation, as Kraft has recorded, was still thriving in the 1970s.[81] This separation was also gendered. As Herbert D. Benington, one of the managers of SAGE, later narrated, "women turned out to be very good for the administrative programs. One reason is that these people tend to be fastidious—they worry how all the details fit together while still keeping the big picture in mind. I don't want to sound sexist, but one of our strongest groups had 80 percent women in it; they were doing the right kind of thing. The mathematicians were needed for some of the more complex applications."[82] Not accidentally, the SDC was spun off from the System Training Program, a group comprised of RAND psychologists focused on producing more effective groups.[83]

Buttressing this hierarchy was a strict system of control, "tools of a very complex nature" that did not survive SAGE. As Benington explains, these tools enabled managers to track and punish coders: "You could assign an individual a job, you could control the data that that individual had access to, you could control when that individual's program operated, and you could find out if that individual was playing the game wrong and punish that person. So we had a whole set of tools for design, for controlling of the team, for controlling of the data, and for testing the programs that were really quite advanced."[84] Because of this system of control, Benington viewed symbolic addressing and other moves to automate programming as "dangerous because they couldn't be well-disciplined." However, although automatic programming has been linked to empowerment, it has also led to the more thorough (because subtle and internalized) disciplining of programmers, which simultaneously empowers and disempowers programmers.

Indeed, this overt system of control and punishment was replaced by a "softer" system of structured programming that makes source code source. As Mahoney has argued, structured programming emerged as a "means both of quality control and of disciplining programmers, methods of cost accounting and estimation, methods of verification and validation, techniques of quality assurance."[85] Kraft targets structured programming as de-skilling: through it, programming was turned from a craft to an industrialized practice in which workers were reduced to interchangeable detail workers.[86] Structured programming limits the logical procedures coders can use and insists that the program consist of small modular units, which can be called from the main program. Structured programming (also generally known as "good programming" when I was growing up) hides, and thus secures, the machine. It focuses on and enables abstraction—and abstraction from the specific uses of and for the machine—thereby turning programming from a numerical- to a problem-based task.

Not surprisingly, having little to no contact with the actual machine enhances one's ability to think abstractly rather than numerically. Edsger Dijkstra, whose famous condemnation of "goto" statements has encapsulated to many the fundamental tenets of structured programming, believes that he was able to "pioneer" structured programming precisely because he began his programming career by coding for ghosts: for machines that did not yet exist.[87] In "Go To Statement Considered Harmful," Dijkstra argues, "the quality of programmers is a decreasing function of the density of go to statements in the programs they produce" because goto statements work against the fundamental tenet of what Dijkstra considered to be good programming, namely, the necessity to "shorten the conceptual gap between the static program and the dynamic process, to make the correspondence between the program (spread out in text space) and the process (spread out in time) as trivial as possible."[88] This is important because, if a program suddenly halts because of a bug, gotos (statements that tell a program to go to a specific line if a condition is met) make it difficult to find the place in the program that corresponds to the buggy code. Gotos make difficult the conflation of instruction with its product—the reduction of process to command—that grounds the emergence of software as a concrete entity and commodity. That is, gotos make it difficult for the source program to act as a legible source.[89] As this example makes clear, structured programming moves away from issues of program efficiency—the time it takes to run a program—and more toward the problem of minimizing all the costs involved in producing and maintaining large programs. This move also makes programming an "art." As Dijkstra argues in his letter justifying structured programming, "it is becoming most urgent to stop to consider programming primarily as the minimization of cost/performance ratio. We should recognize that already now programming is much more an intellectual challenge: the art of programming is the art of organizing complexity, of mastering multitude and avoiding its bastard chaos as effectively as possible."[90] Again, this depends on making "the structure of the program text [reflect] the structure of the computation."[91] It means moving away from assembly and other languages that routinely offer bizarre exits and self-modifying code to languages that feature clear and well-documented repetitions (while . . . do . . .) that end in one clear place, that return control to the main program.

Structured programming languages "save" programmers from themselves by providing good security, where security means secure from the programmer (increasingly, "securing" the machine means making sure programmers cannot access or write over key systems).[92] Indeed, structured programming, which emphasizes programming as a problem of flow, is giving way to data abstraction, which views programming as a problem of interrelated objects, and hides far more than the machine. Data abstraction depends on information hiding, on the nonreflection of changeable facts in software. As John V. Guttag, a "pioneer" in data abstraction explains, data abstraction is all about forgetting, about hiding information about how a type is implemented behind an interface.[93] Rather than "polluting" a program by enabling invisible lines of contact between supposedly independent modules, data abstraction presents a clean or "beautiful" interface by confining specificities, and by reducing the knowledge and power of the programmer. Knowledge, Guttag insists, is dangerous: "'Drink deep, or taste not the Pierian Spring,' is not necessarily good advice. Knowing too much is no better, and often worse, than knowing too little. People cannot assimilate very much information. Any programming method or approach that assumes that people will understand a lot is highly risky."[94] Abstraction—the "erasure of difference in the service of likeness or equality"—also erases, or "forgets," knowledge, rendering it, like the machine, ghostly.[95]

Thus abstraction both empowers the programmer and insists on his/her ignorance—the dream of a sovereign subject who knows and commands is constantly undone. Because abstraction exists "in the mind of the programmer," abstraction gives programmers new creative abilities. Computer scientist David Eck argues, "every programming language defines a virtual machine, for which it is the machine language. Designers of programming languages are creating computing machines as surely as the engineer who works in silicon and copper, but without the limitations imposed by materials and manufacturing technology."[96] However, this abstraction—this move away from the machine specificities—hands over, in its virtual separation of machine into software and hardware, the act of programming to the machine itself. Mildred Koss scoffed at the early notion of computers as brains because "they couldn't think in the way a human thinks, but had to be given a set of step-by-step machine instructions to be executed before they could provide answers to a specific problem"—at that time software was not considered to be an independent object.[97] The current status of software as a commodity, despite the nonrivalrous nature of "instructions," indicates the triumph of the software industry, an industry that first struggled not only financially but also conceptually to define its product. The rise of software depends both on historical events, such as IBM's unbundling of its services from its products, and on abstractions enabled by higher-level languages. Guttag's insistence on the unreliability and incapability of human beings to understand underscores the cost of such an abstraction. Abstraction is the computer's game, as is programming in the strictest and newest sense of the word: with "data-driven" programming, for instance, machine learning/artificial intelligence (computers as source of source code) has become mainstream.

Importantly, this stratification and disciplining of labor has a much longer history: human computing itself, as David Grier has documented, moved from an art to a routinized procedure through a separation of planners from calculators.[98] Whereas the mathematician Alexis-Claude Clairaut called on two of his colleagues/friends, Joseph Lalande, Nicole-Reine Lapaute, in 1757 to calculate the date of Halley's comet's 1758 return, Gaspard Clair François Marie Riche de Prony, director of the Bureau du Cadastre, devised a system of intellectual labor to calculate metric tables in 1791. Not accidentally, the tables were part of a revolutionary governmental project: the move to the metric system by the National Assembly in order to gain control of the French economy.[99] De Prony, inspired by Adam Smith, divided the group into manual workers (unemployed pre-Revolutionary wig makers or servants who had basic arithmetic skills) and planners (experienced computers who planned the calculation). This system in turn inspired Charles Babbage's difference and analytic engines, in which the engines would replace the manual workers: according to Grier, de Prony's system showed Babbage that "the division of labor was not restricted to physical work but could be applied to 'some of the sublimest investigations of the human mind,' including the work of calculation."[100] This routinized calculation was not smoothly adopted; for a long time within the United States, such a model was resisted and, even during World War I, computers were graduate students and young assistant professors. In order to produce calculations necessary for governmental projects (such as eugenics, census, navigation, weapons, etc.) in the twentieth century, however, mass computation became the norm.

The U.S. wholesale embrace of mass calculation also coincides with a governmental project. Begun during the Great Depression as a way to put unemployed high school graduates to work, the Work Progress Administration's (WPA) Math Tables Project (MTP) produced some of the finest error-free tables in the world.[101] Indeed, it was not until the Roosevelt administration and the New Deal that the United States became seriously involved in producing mathematical tables. Since it was a WPA project, many established academics refused to be involved with it. To gain credibility, those in charge (themselves "less desirable" or unconventional PhDs) were determined to produce the most accurate tables possible. Gertrude Blanch, who ran the program with Milton Abramowitz, insists that most of the people they hired were qualified.[102] In contrast, Ida Rhodes, another PhD hired by the MTP, claims: "[Most] of the people

[who] came to us really knew nothing at all about mathematics or [even] arithmetic. Gertrude Blanch says that they were all High School graduates, and they may have been. I never checked on that. But if they were, very few of them had remembered anything about the arithmetic or the algebra or whatever mathematics they had [studied]."[103] By the end, however, they were transformed. According to Rhodes, Blanch performed miracles, "welding a malnourished, dispirited crew of people, coming from [the] Welfare Rolls, [into] a group that Leslie J. Comrie said was the 'mightiest computing team the world had ever seen.'"[104] To Rhodes, the social work involved in this project—"[salutary benefit conferred on] the spirit of those people [by] raising them from abject and self-despising people into a team that [acquired] a magnificent esprit de corps"—has been overlooked.[105] As Rhodes's rhetoric indicates, this was a patronizing if admirable project, run by "saints." Rhodes, herself partially deaf, would become an advocate for including physically challenged people in programming work. (Blanch interestingly had a more edgy view of sainthood. Describing Rhodes, she remarked, "if there are saints on earth, she's one of them. Saints may be difficult to live with but . . . it's nice to have a few around").[106]

This saintly salutary work comprised dividing the group into four categories, listed in ascending ability—the adders, the multipliers, the dividers, and the checkers—and creating worksheets so that "people who knew nothing about mathematics could [do advanced functions] by just following one step at a time."[107] The flawlessness of these tables stemmed both from these worksheets, created by Blanch, and from the degree to which these tables were checked (the Bessel function, for instance, was checked more than twenty-two times). Since the goal of the project was to keep these people busy, as well as to produce tables, accuracy was stressed over expediency and over sophistication of numerical techniques. Accuracy, according to Rhodes, became an obsession."[108]

Not surprisingly, though, the MTP computers were sometimes suspicious of their oversight. Rhodes relates, "we had impressed upon our workers over and over and over again that we were not watching them. We were not counting their output." Rather, "the only thing we asked of them is complete accuracy." This accuracy was also inscribed in the worksheets themselves in a nontransparent, repetitive manner. Rhodes and Blanch created worksheets, "in which every operation had to be done at least twice" and in which this duplicity was hidden. Rhodes explains, "for example, if we added a and b we wouldn't immediately say: add b and a. But some time later we saw to it that b got added to a, and we had arrows connecting the answers saying that these two answers should agree to, say one or two [units in] the last place. If they did not get such an agreement, then they were to [erase the pertinent portion] and [re-compute it]."[109] Again, the fact that these tables were largely unnecessary—and hence not time-sensitive—made this emphasis on accuracy over timeliness possible.

According to Rhodes, only two girls did not internalize the accuracy-ethic and cheated.[110] Rhodes revealingly narrates the dishonesty of the "colored" girl who joined the group after claiming that she was being discriminated against in another project:

[Being] a softy, [I swallowed her story.] I should have checked with [her] boss and found out why she was not liked. But I didn't. And so I asked Gertrude's permission and she said, "All right, let's give her a chance." [And] she started working for us.

Well, she hadn't been with us long enough apparently to absorb that feeling of accuracy, although, of course, we also gave her the [same] lecture that we gave everybody else. She must have thought that the more she produces, the more we will think of her and the more anxious we will be to keep her. [Her checker] reported to us that the girl was a whiz, she handed in many more sheets than anyone else; and I began to feel very proud of myself, thinking, oh, I got [me a] good girl, working so hard.

You see, all that the [checker did was to examine the values, connected by the] arrows and if they agreed within one or two [units,] he was satisfied. In her case he once mentioned, "It's remarkable, they agree to the very last place." That should have given me an idea, but I was too busy with other things. Well, one evening Gertrude and I sat down to do our regular job of checking the sheets, and [when] we got [to] hers, [no values] differenced, absolutely nothing differenced. That was something we couldn't believe. How could [they] not difference? The arrows showed perfect agreement—too perfect, as a matter of fact.

Well, lots of things can happen. First of all, the formula can be wrong. [Or we] could have made a mistake [in breaking down] the formula [while preparing] the worksheet. [Or] we could have made a mistake in [a sign.] We could have made a mistake in a constant. It happened to be my worksheet, so I checked [it] over: no mistake there. [She had to] copy certain information from other Tables. Maybe [I] gave her the wrong tables. [An examination showed] that she copied the correct Tables. What else could have happened? The point [is] that we were so innocent and so trusting, it never occurred to us that what really happened [could have occurred.] What had happened was that she would get the first answer, and then when she got to [it] the second time — where the arrows showed that they had to agree — and [they] didn't agree, she merely erased the [second] answer and copied down the first [one.] We found that out [when] Gertrude and I recomputed all her sheets.[111]

This remarkable story reveals the contradictions in this disciplinary system: although Rhodes denies that they judged performance by speed, she thinks she got herself a "good girl" when the "colored girl" performs quickly. Also, although math presumably requires some intellectual labor, intelligence is condemned. The "colored girl"'s ability to figure out the system, the algorithm, is denounced as cheating, and the managers' faith in their own nontransparent plans described as "trusting." These worksheets were an early form of programming: a breakdown of a complex operation into sequence of simple operations that depends on accurate and single-minded calculation. As this example makes clear, such programming depended on mind-numbingly repetitive operations by the "dumb" and the downtrodden, whose inept or deceitful actions could disrupt the task at hand. Modern computing replaces these with vacuum tubes and transistors.

As Alan Turing contended, "the class of problems capable of solution by the machine can be defined fairly specifically . . . [namely] those problems which can be solved by human clerical labour, working to fixed rules, and without understanding."[112]

Source code become "thing"—the erasure of execution—follows from the mechanization of these power relations, the reworking of subject-object relations through automation as both empowerment and enslavement and through repetition as both mastery and hell. Embedded within the notion of instruction as source and the drive to automate computing—relentlessly haunting them—is a constantly repeated narrative of liberation and empowerment, wizards and (ex-)slaves.

Automation as Sourcery

Automatic programming, what we could call programming today, reveals the extent to which automation and the history of programming cannot be considered a simple deskilling (Kraft's argument) or a march toward greater human power. Rather, through automation, expertise is both created and called into question: it is something that coders did not simply fear, but also appreciated and drove.

Automatic programming arose from a desire to reuse code and to recruit the computer into its own operation—essentially, to transform singular instructions into a language a computer could write. As Koss, an early UNIVAC programmer, explains:

> Writing machine code involved several tedious steps—breaking down a process into discrete instructions, assigning specific memory locations to all the commands, and managing the I/O buffers. After following these steps to implement mathematical routines, a sub-routine library, and sorting programs, our task was to look at the larger programming process. We needed to understand how we might reuse tested code and have the machine help in programming. As we programmed, we examined the process and tried to think of ways to abstract these steps to incorporate them into higher-level language. This led to the development of interpreters, assemblers, compilers, and generators—programs designed to operate on or produce other programs, that is, automatic programming.[113]

Automatic programming is an abstraction that allows the production of computer-enabled human-readable code—key to the commodification and materialization of software and to the emergence of higher-level programming languages.

Higher-level programming languages, unlike assembly language, explode one's instructions and enable one to forget the machine. In them, simple operations often call a function, making it a metonymic language par excellence. These languages also place everyone in the position of the planner, without the knowledge of the coder. They enable one to run a program on more than one machine—a property now assumed to be a "natural" property of software ("direct programming" led to a unique configuration of cables; early machine language could be iterable but only on the same machine—assuming, of course, no engineering faults or failures). In order to emerge as a language or a source, software and the "languages" on which it relies had to become iterable. With programming languages, the product of programming would no longer be a running machine but rather this thing called software—something theoretically (if not practically) iterable, repeatable, reusable, no matter who wrote it or what machine it was destined for; something that inscribes the absence of both the programmer and the machine in its so-called writing.[114] Programming languages enabled the separation of instruction from machine, of imperative from action, a move that fostered the change in the name of source code itself, from "pseudo" to "source." Pseudocode intriguingly stood both for the code as language and for the code as program (i.e., source code). The manual for UNIVAC's A-2 compiler, for instance, defines pseudocode as "computer words other than the machine (C-10) code, design [sic] with regard to facilitating communications between programmer and computer. Since a pseudo-code cannot be directly executed by the computer, there must be programmed a modification, interpretation or translation routine which converts the pseudo-codes to machine instruction and routines."[115] Pseudocode, which enables one to move away from machine specificity, is called "information"—what later would become a ghostly immaterial substance—rather than code.

According to received wisdom, these first attempts to automate programming—the "pseudo"—were resisted by "real" programmers.[116] John Backus, developer of FORTRAN, claims that early machine language programmers were engaged in a "black art"; they had a "chauvinistic pride in their frontiersmanship and a corresponding conservatism, so many programmers of the freewheeling 1950s began to regard themselves as members of a priesthood guarding skills and mysteries far too complex for ordinary mortals."[117] Koss similarly argues, "without these higher-level languages and processes . . . , which democratized problem solving with the computer, I believe programming would have remained in the hands of a relatively small number of technically oriented software writers using machine code, who would have been essentially the high priests of computing."[118]

This story of a "manly" struggle against automatic programming resonates with narratives of mechanical computing itself as "feminizing" numerical analysis. Whirlwind team member Bob Everett offers the following summary of a tale describing two different ways of approaching automatic computing, which was told at Aiken's mid-1940s meeting: "One was the woman who gets married, and that's fine, and she looks ahead to a life-time of three meals a day, 365 days a year, and dishes to wash after each one of them. Her husband brings her home from the honeymoon, and she discovers he's bought her an automatic dishwasher. That's one way. The other way is the guy who decides to climb a mountain, and he buys all the rope, pitons, and one thing and another, and he goes to the mountain and finds that somebody has built a funicular railway."[119] According to this description, automatic computing is feminine or emasculating: an escape from domestic drudgery or the automation of a properly masculine enterprise. Thus, it is not just the introduction of automatic programming that inspired narratives of masculine expertise under siege, but also the introduction of—or, more properly, the appreciation of—the (automatic) computer.

In a related manner, Hopper (and perhaps only Hopper) experienced the U.S. Navy, in particular her initial training as a thirty-seven-year-old woman, as "the most complete freedom I'd ever had." Whereas her younger counterparts rebelled "against the uniforms and the regulations," she embraced the Navy's strict structure as a release from domestic duties. As she relates, "All of a sudden I didn't have to decide anything, it was all settled. I didn't even have to bother to decide what I was going to wear in the morning, it was there. I just picked it up and put it on. So for me all of a sudden I was relieved of all minor decisions. . . . I didn't even have to figure out what I was going to cook for dinner." The difficulties of domestic life and sacrifice during World War II colored Hopper's enthusiasm, since "housekeeping had gotten to be quite a chore by then to figure out how much meat you could have and could you give dad some sugar 'cause he loved it and you might have some extra points. That's when I learned to drink most of my drinks without any sugar in them so that dad could have it. And we had very little gasoline and we had to have a car and you had to plan every trip very carefully. Well, all of a sudden I'm in midshipmen's school and all of a sudden you don't have to do any of it.[120] Importantly, though, this release was also an insertion into a well-defined system, in which one both gave and received commands. When a *Voice of America* interviewer asked, "You are supposed to command, but also to conform and obey. How do you come to terms with those two extremes?" Hopper replied, "The essential basic principle of the Navy is leadership. And leadership is a two-way street. It is loyalty up and loyalty down. Respect your superior, keep him informed of what you are doing, and take care of your crew. That is everyone's responsibility."[121]

Automatic programming, seen as freeing oneself from both drudgery and knowledge, thus calls into question the simple narrative of it as dispersing a reluctant "priesthood" of machine programmers. This narrative of resistance assumes that programmers naturally enjoyed tedious and repetitive numerical tasks and as well as developing singular solutions for their clients. The "mastery" of computing can easily be understood as "suffering." Indeed, Hopper called her early days with the Harvard Mark 1 her "sufferings" and argued, "experienced programmers are always anxious to make the computer carry out as much routine work as they can."[122] Harry Reed, an early ENIAC programmer, relays, "the whole idea of computing with the ENIAC was a sort of *hair-shirt* kind of thing. Programming for the computer, whatever it was supposed to be, was a redemptive experience—one was supposed to *suffer* to do it." According to Reed, programmers were actively trying to convince people to write small programs for themselves. In the 1970s, he "actually had to take my Division and sit everybody down who hadn't taken a course in FORTRAN, because, by God, they were going to write their own programs now. We weren't going to get computer specialists to write simple little programs that they should have been writing.[123] Also, the first programmers were the first writers of reusable subroutines. Holberton, for instance, developed the first SORT generator to save her colleagues' time, "I felt for all the work that Betty Jean and I had done on sorting methods, it was a shame for people to have to sit down and re-do and re-code that same thing even though they could use the books to do it, if it could be done by a machine. And that's the reason, and it only took six months to program the thing. That's six more months."[124] Thus, rather than programmers circling the wagons to protect their positions, it would seem that many programmers themselves welcomed and contributed to the success of automatic programming.

As well, since programmers were in incredible demand in the 1950s through the 1960s, the need to create boundaries to protect jobs seems odd. Although compilers and interpreters may not have been accepted immediately, especially by those already trained in machine programming, the resistance may have stemmed more from the work environment than from personal arrogance. Coders were under great pressure to be as efficient as possible. As Holberton and Bartik relay in a 1973 interview, early coders often developed a persecution complex, because machine time was the most important and expensive thing:

BARTIK: The worst sin that you could commit was to waste that machine time. So that we really became paranoid.
HOLBERTON: Mhm. Efficiency.
BARTIK: We thought everybody was after us.
TROPP: [Laughter]
BARTIK: For our inefficiency.
HOLBERTON: You wasted one add time, you were being inefficient.
BARTIK: So it was fine for us to struggle for two days to cut off the slightest amount on that machine.[125]

Compilers were arguably accepted because the demand for programmers meant a loss in quality (an ever widening recruitment)—programming efficiently in machine language therefore became a mark of expertise. In this sense, the introduction of automatic programming, which set a certain standard of machine efficiency, helped to produce the priesthood it was supposedly displacing.

Corporate and academic customers, for whom programmers were orders of magnitude cheaper per hour than computers, do seem to have resisted automatic programming. Jean Sammet, an early female programmer, relates, in her influential *Programming Languages: History and Fundamentals*, that customers objected to compilers on the ground that they "could not turn out object code as good as their best programmers. A significant selling campaign to push the advantages of such systems was underway at that time, with the spearhead being carried for the numerical scientific languages (i.e., FORTRAN) and for 'English-language-like' business data-processing languages by

Remington Rand (and Dr. Grace Hopper in particular)."[126] This selling campaign not only pushed higher-level languages (by devaluing humanly produced programs), it also pushed new hardware: to run these programs, one needed more powerful machines. The government's insistence on standardization, most evident in the development and widespread use of COBOL, itself a language designed to open up programming to a wider range of people, fostered the general acceptance of higher-level languages, which again were theoretically, if not always practically, machine independent or iterable. The hardware-upgrade cycle was normalized in the name of saving programming time.

This "selling campaign" led to what many have heralded as the democratization of programming, the opening of the so-called priesthood of programmers. In Sammet's view, this was a partial revolution

> in the way in which computer installations were run because it became not only possible, but quite practical to have engineers, scientists, and other people actually programming their own problems without the intermediary of a professional programmer. Thus the conflict of the open versus closed shop became a very heated one, often centering [on] the use of FORTRAN as the key illustration for both sides. This should not be interpreted as saying that all people with scientific numerical problems to solve immediately sat down to learn FORTRAN; this is clearly not true but such a significant number of them did that it has had a major impact on the entire computer industry. One of the subsidiary side effects of FORTRAN was the introduction of FORTRAN Monitor System [IB60]. This made the computer installation much more efficient by requiring less operator intervention for the running of the vast number of FORTRAN (as well as machine language) programs.[127]

The democratization or "opening" of computing, which gives the term *open* in *open source* a different resonance, would mean the potential spread of computing to those with scientific numerical problems to solve and the displacement of human operators by operating systems. But the language of priests and wizards has hardly faded and scientists have always been involved with computing, even though computing has not always been considered to be a worthy scientific pursuit. The history of computing is littered with moments of "computer liberation" that are also moments of greater obfuscation.[128]

Higher level programming languages—automatic programming—may have been sold as offering the programmer more and easier control, but they also necessitated blackboxing even more the operations of the machine they supposedly instructed. Democratization did not displace professional programmers but rather buttressed their position as professionals by paradoxically decreasing their real power over their machines, by generalizing the engineering concept of information.

So what are we to do with these contradictions and ambiguities? As should be clear by now, these many contradictions riddling the development of automatic programming were key to its development, for the automation of computing is both an acquisition of greater control and freedom, and a fundamental loss of them. The narrative of the "opening" of programming reveals the tension at the heart of programming and control systems: are they control systems or servomechanisms (Norbert Wiener's initial name for them)? Is programming a clerical activity or an act of Hobbesian mastery? Given that the machine takes care of "programming proper"—the sequence of events during execution—is programming programming at all? What is after all compacted in the coinciding changes in the titles of "operators" to "programmers" and of "mathematicians" to "programmers"? The notion of the priesthood of programming erases this tension, making programming always already the object of jealous guardianship, and erasing programming's clerical underpinnings.[129]

Programming in the 1950s does seem to have been fun and fairly gender balanced, in part because it was so new and in part because it was not as lucrative as hardware design or even sales: the profession was gender neutral in hiring if not pay because it was not yet a profession.[130] The "ENIAC girls" were first hired as subprofessionals, and some had to acquire more qualifications in order to retain their positions. As many female programmers quit to have children or get married, men (and compilers) took their increasingly lucrative positions. Programming's clerical and arguably feminine underpinnings—both in terms of personnel and of command structure—became buried as programming sought to become an engineering and academic field in its own right.[131] Democratization did not displace professional programmers but rather buttressed their position as professionals by paradoxically decreasing their real power over their machines. It also, however, made programming more pleasurable.

Causal Pleasure

The distinction between programmers and users is gradually eroding. With higher-level languages, programmers are becoming more like simple users. Crucially, though, the gradual demotion of programmers has been offset by the power and pleasure of programming. To program in a higher-level language is to enter a magical world—it is to enter a world of logos, in which one's code faithfully represents one's intentions, albeit through its blind repetition rather than its "living" status.[132] Edwards argues, "programming can produce strong sensations of power and control" because the computer produces an internally consistent if externally incomplete microworld, a "simulated world, entirely within the machine itself, that does not depend on instrumental effectiveness. That is, where most tools produce effects on a wider world of which they are only a part, the computer contains its own worlds in miniature. . . . In the microworld, as in children's make-believe, the power of the programmer is absolute."[133] Joseph Weizenbaum, MIT professor, creator of ELIZA (an early program that imitated a Rogerian therapist) and member of the famed MIT AI (Artifical Intelligence) lab, similarly contends:

The computer programmer . . . is a creator of universes for which he alone is the lawgiver. So, of course, is the designer of any game. But universes of virtually unlimited complexity can be created in the form of computer programs. Moreover, and this is a crucial point, systems so formulated and elaborated *act out* their programmed scripts. They compliantly obey their laws and vividly exhibit their obedient behavior. No playwright, no stage director, no emperor, however powerful, has ever exercised such absolute authority to arrange a stage or a field of battle and to command such unswervingly dutiful actors or troops.[134]

The progression from playwright to stage director to emperor is telling: programming languages, like neoliberal economics, model the world as a "game."[135] To return to the notion of "code is law," programming languages establish the programmer as a sovereign subject, for whom there is no difference between command given and command completed. As a lawgiver more powerful than a playwright or emperor, the programmer can "say" "let there be light" and there is light. Iterability produces both language and subject. Importantly, Weizenbaum views the making performative or automatically executable of words as the imposition of instrumental reason, inseparable from the process of "enlightenment" critiqued by the Frankfurt school.[136] Instrumental reason, he argues, "has made out of words a fetish surrounded by black magic. And only the magicians have the rights of the initiated. Only they can say what words mean. And they play with words and they deceive us."[137]

Programming languages offer the lure of visibility, readability, logical if magical cause and effect. As Brooks argues, "one types the correct incantation on the keyboard, and a display screen comes to life, showing things that never were nor could be."[138] One's word creates something living. Consider this ubiquitous "hello world" program written in C++ ("hello world" is usually the first program a person will write):

```
// this program spits out "hello world"

#include <iostream.h>

int main ()

{

    cout << "Hello World!";

    return 0;

}
```

The first line is a comment line, explaining to the human reader that this program spits out "Hello World!." The next line directs the compiler's preprocessor to include iostream.h, a standard file to deal with input and output to be used later. The third line, "int main ()," begins the main function of the program; "cout << 'Hello World!';" prints "Hello World!" to the screen ("cout" is defined in iostream.h); "return 0" terminates the main function and causes the program to return a 0 if it has run correctly.

Although not immediately comprehensible to someone not versed in C++, this program nonetheless seems to make some sense, and seems to be readable. It comprises a series of imperatives and declaratives that the computer presumably understands and obeys. When it runs, it follows one's commands and displays "Hello World!."

It is no accident that "hello world" is the first program one learns because it is easy, demonstrating that we can produce results immediately. This ease, according to Weizenbaum, is what makes programming so seductive and dangerous:

> It happens that programming is a relatively easy craft to learn. . . . And because programming is almost immediately rewarding, that is, because a computer very quickly begins to behave somewhat in the way the programmer intends it to, programming is very seductive, especially for beginners. Moreover, it appeals most to precisely those who do not yet have sufficient maturity to tolerate long delays between an effort to achieve something and the appearance of concrete evidence of success. Immature students are therefore easily misled into believing that they have truly mastered a craft of immense power and of great importance when, in fact, they have learned only its rudiments and nothing substantive at all.[139]

The seeming ease of programming hides a greater difficulty—executability leads to unforeseen circumstances, unforeseen or buggy repetitions. Programming offers a power that, Weizenbaum argues, corrupts as any power does.[140] *What corrupts, Weizenbaum goes on to explain, however, is not simply ease, but also this combination of ease and difficulty.* Weizenbaum argues that programming creates a new mental disorder: the compulsion to program, which he argues hackers, who "hack code" rather than "work," suffer from (although he does note that not all hackers are compulsive programmers).[141]

To explain this addiction, Weizenbaum explains the parallels between "the magical world of the gambler" and the magical world of the hacker—both entail megalomania and fantasies of omnipotence, as well as a "pleasureless drive for reassurance."[142] Like gambling, programming can be compulsive because it both rewards and challenges the programmer. It is driven by "two apparently opposing facts: first, he knows that he can make the computer do anything he wants it to do; and second, the computer constantly displays undeniable evidence of his failures to him. It reproaches him. There is no escaping this bind. The engineer can resign himself to the truth that there are some things he doesn't know. But the programmer moves in a world entirely of his own making. The computer challenges his power, not his knowledge."[143] According to Weizenbaum, because programming engages power rather than truth, it can induce a paranoid megalomania in the programmer.[144] Because this knowledge is never enough, because a new bug always emerges, because an unforeseen wrinkle causes divergent unexpected behavior, the hacker can never stop. Every error seems correctable; every error points to the hacker's lack of foresight; every error leads to another. Thus, unlike the "useful programmer," who "works" by solving the problem at hand and carefully documents his code, the hacker aimlessly hacks code: programming becomes a technique, a game without a goal and thus without an end. Hackers' skills are thus "disembodied" and this disembodiment transforms their physical appearance: Weizenbaum describes them as "bright young men of disheveled appearance, often with sunken glowing eyes . . . sitting at computer consoles, their arms tensed and waiting to fire their fingers, already poised to strike, at the buttons and keys on which their attention seems to be as riveted as a gamer's on the rolling dice.[145]

Although Weizenbaum is quick to pathologize hackers as pleasureless pitiful creatures, hackers themselves emphasize programming as pleasurable—and their lack of "usefulness" can actually be what is most productive and promising about programming. Linus Torvalds, for instance, argues that he, as an eternal grad student, decided to build the Linux operating system core just "for fun." Torvalds further views the decisions programming demands as rescuing programming from becoming tedious. "Blind obedience on its own, while initially fascinating," he writes, "obviously does not make for a very likable companion. In fact, that part gets boring fairly quickly. What makes programming so engaging is that, while you can make the computer do what you want, you have to figure out *how*."[146] Richard Stallman, who fits Weizenbaum's description of a hacker (and who was in the AI lab, probably building those indispensable functions) likewise emphasizes the pleasure, but more important the "freedom" and "freeness" associated with programming—something that stems from programming as not simply the production of a commercial (or contained) product. Hacking reveals the extent to which source code can become a fetish: something endless that always leads us pleasurably, as well as anxiously, astray.

Source Code as Fetish

Source code as source means that software functions as an axiom, as "a self-evident proposition requiring no formal demonstration to prove its truth, but received and assented to as soon as it is mentioned."[147] In other words, whether or not source code is only a source after the fact or whether or not software can be physically separated from hardware,[148] software is always posited as already existing, as the self-evident ground or source of our interfaces. Software is axiomatic. As a first principle, it fastens in place a certain neoliberal logic of cause and effect, based on the erasure of execution and the privileging of programming that bleeds elsewhere and stems from elsewhere as well.[149] As an axiomatic, it, as Gilles Deleuze and Félix Guattari argue, artificially limits decodings.[150] It temporarily limits what can be decoded, put into motion, by setting up an artificial limit—the artificial limit of programmability—that seeks to separate information from entropy, by designating some entropy information and other "non-intentional" entropy noise. Programmability, discrete computation, depends on the disciplining of hardware and programmers, and the desire for a programmable axiomatic code. Code, however, is a medium in the full sense of the word. As a medium, it channels the ghost that we imagine runs the machine—that we see as we don't see—when we gaze at our screen's ghostly images.

Understood this way, source code is a fetish. According to the OED, a fetish was originally an ornament or charm worshipped by "primitive peoples . . . on account of its supposed inherent magical powers."[151] The term *fetisso* stemmed from the trade of small wares and magic charms between the Portuguese merchants and West Africans; Charles de Brosses coined the term *fetishism* to describe "primitive religions" in 1757. According to William Pietz, Enlightenment thinkers viewed fetishism as a "false causal reasoning about physical nature" that became "the definitive mistake of the pre-enlightened mind: it superstitiously attributed intentional purpose and desire to material entities of the natural world, while allowing social action to be determined by the . . . wills of contingently personified things, which were, in truth, merely the externalized material sites fixing people's own capricious libidinal imaginings."[152] That is, fetishism, as "primitive causal thinking," derived causality from "things"—in all the richness of this concept—rather than from reason:

> Failing to distinguish the intentionless natural world known to scientific reason and motivated by practical material concerns, the savage (so it was argued) superstitiously assumed the existence of a unified causal field for personal actions and physical events, thereby positing reality as subject to animate powers whose purposes could be divined and influenced. Specifically, humanity's belief in gods and supernatural powers (that is, humanity's unenlightenment) was theorized in terms of prescientific peoples' substitution of imaginary personifications for the unknown physical causes of future events over which people had no control and which they regarded with fear and anxiety.[153]

A fetish allows one to visualize what is unknown—to substitute images for causes. Fetishes allow the human mind both too much and not enough control by establishing a "unified causal field" that encompasses both personal actions and physical events. Fetishes enable a semblance of control over future events—a possibility of influence, if not an airtight programmability—that itself relies on distorting real social relations into material givens.

This notion of fetish as false causality has been most important to Karl Marx's diagnosis of capital as fetish. Marx famously argued:

> the commodity-form . . . is nothing but the determined social relation between men themselves which assumes here, for them, the phantasmagoric form of a relation between things. In order, therefore, to find an analogy we must take a flight into the misty realm of religion. There the products of the human head appear as autonomous figures endowed with a life of their own, which enter into relations both with each other and with the human race. So it is in the world of commodities with the products of men's hands. I call this the . . . fetishism.[154]

The capitalist thus confuses social relations and the labor activities of real individuals with capital and its seemingly magical ability to reproduce. For, "it is in interest-bearing capital . . . that capital finds its most objectified form, its pure fetish form. . . . Capital—as an entity—appears here as an independent source of value; a something that creates value in the same way as land [produces] rent, and labor wages."[155] Both these definitions of fetish also highlight the relation between things and men: men and things are not separate, but rather speak with and to one another. That is, things are not simply objects that exist outside the human mind, but are rather tied to events, to the timing of events.

The parallel to source code seems obvious: we "primitive folk" worship source code as a magical entity—as a source of causality—when in truth the power lies elsewhere, most importantly, in social and machinic relations. If code is performative, its effectiveness relies on human and machinic rituals. Intriguingly though, in this parallel, Enlightenment thinking—a belief that knowing leads to control, to a release from tutelage—is not the "solution" to the fetish, but, rather, what grounds it, for source code historically has been portrayed as the solution to wizards and other myths of programming: machine code provokes mystery and submission; source code enables understanding and thus institutes rational thought and freedom. Knowledge, according to Weizenbaum, sustains the hacker's aimless actions. To offer a more current example of this logic than the FORTRAN one cited earlier, Richard Stallman, in his critique of nonfree software, has argued that an executable program "is a mysterious bunch of numbers. What it does is secret."[156] Against this magical execution, source code supposedly enables an understanding and a freedom—the ability to map and know the workings of the machine, but, again, only through a magical erasure of the gap between source and execution, an erasure of execution itself. If we consider source code as fetish, the fact that source code has hardly deprived programmers of their priestlike/wizard status makes complete sense. If anything, such a notion of programmers as superhuman has been disseminated ever more and the history of computing—from direct manipulation to hypertext—has been littered by various "liberations."

But clearly, source code can do and be things: it can be interpreted or compiled; it can be rendered into machine-readable commands that are then executed. Source code is also read by humans and is written by humans for humans and is thus the source of some understanding. Although Ellen Ullman and many others have argued, "a computer program has only one meaning: what it does. It isn't a text for an academic to read. Its entire meaning is its function," source code must be able to function, even if it does not function—that is, even if it is never executed.[157] Source code's readability is not simply due to comments that are embedded in the source code, but also due to English-based commands and programming styles designed for comprehensibility. This readability is not just for "other programmers." When programming, one must be able to read one's own program—to follow its logic and to predict its outcome, whether or not this outcome coincides with one's prediction.

This notion of source code as readable—as creating some outcome regardless of its machinic execution—underlies "codework" and other creative projects. The Internet artist Mez, for instance, has created a language called mezangelle that incorporates formal code and informal speech. Mez's poetry deliberately plays with programming syntax, producing language that cannot be executed, but nonetheless draws on the conventions of programming language to signify.[158] Codework, however, can also work entirely within an existing programming language. Graham Harwood's perl poem, for example, translates William Blake's nineteenth-century poem "London" into London. pl, a script that contains within it an algorithm to "find and calculate the gross lung-capacity of the children screaming from 1792 to the present."[159] Regardless of whether or not it can execute, code can be—must be—worked into something meaningful. Source code, in other words, may be the source of things other than the machine execution it is "supposed" to engender.

Source code as fetish, understood psychoanalytically, embraces this nonteleological potential of source code, for the fetish is a deviation that does not "end" where it should. It is a genital substitute that gives the fetishist nonreproductive pleasure. It allows the child to combat castration—his inscription within the world of paternal law and order—for both himself and his mother, while at the same time accommodating to his world's larger oedipal structure. It both represses and acknowledges paternal symbolic authority. According to Freud, the fetish, formed the moment the little boy discovers his mother's "lack," is "a substitute for the woman's (mother's) phallus which the little boy once believed in and does not wish to forego."[160] As such, it both fixes a singular event—turning time into space—and enables a logic of repetition that constantly enables this safeguarding. As Pietz argues, "the fetish is always a meaningful fixation of a singular event; it is above all a 'historical' object, the enduring material form and force of an unrepeatable event. This object is 'territorialized' in material space (an earthly matrix), whether in the form of a geographical locality, a marked site on the surface of the human body, or a medium of inscription or configuration defined by some portable or wearable thing."[161] Even though it fixes a singular event, the fetish works only because it can be repeated, but again, what is repeated is both denial and acknowledgment, since the fetish can be "the vehicle both of denying and asseverating the fact of castration."[162] Slavoj Žižek draws on this insight to explain the persistence of the Marxist fetish:

> When individuals use money, they know very well that there is nothing magical about it—that money, in its materiality, is simply an expression of social relations . . . on an everyday level, the individuals know very well that there are relations between people behind the relations between things. The problem is that in their social activity itself, in what they are *doing*, they are *acting* as if money, in its material reality is the immediate embodiment of wealth as such. They are fetishists in practice, not in theory. What they "do not know,"

what they misrecognize, is the fact that in their social reality itself—in the act of commodity exchange—they are guided by the fetishistic illusion.[163]

Fetishists, importantly, know what they are doing—knowledge, again, is not an answer to fetishism, but rather what sustains it. The knowledge that source code offers is no cure for source code fetishism: if anything, this knowledge sustains it. As the next chapter elaborates, the key question thus is not "what do we know?" but rather "what do we do?"

To make explicit the parallels, source code, like the fetish, is a conversion of event into location—time into space—that does affect things, although not necessarily in the manner prescribed. Its effects can be both productive and nonexecutable. Also, in terms of denial and acknowledgment, we know very well that source code in that state and without the intercession of other "layers" is not executable, yet we persist in treating it as so. And it is this glossing over that makes possible the ideological belief in programmability.

Code as fetish means that computer execution deviates from the so-called source, as source program does from programmer. Turing, in response to the objection that computers cannot think because they merely follow human instructions, contends:

> Machines take me by surprise with great frequency. . . . The view that machines cannot give rise to surprises is due, I believe, to a fallacy to which philosophers and mathematicians are particularly subject. This is the assumption that as soon as a fact is presented to a mind all consequences of that fact spring into the mind simultaneously with it. It is a very useful assumption under many circumstances, but one too easily forgets that it is false. A natural consequence of doing so is that one then assumes that there is no virtue in the mere working out of consequences from data and general principles.[164]

This erasure of the vicissitudes of execution coincides with the conflation of data with information, of information with knowledge—the assumption that what is most difficult is the capture, rather than the analysis, of data. This erasure of execution through source code as source creates an intentional authorial subject: the computer, the program, or the user, and this source is treated as the source of meaning. The fact that there is an algorithm, a meaning intended by code (and thus in some way knowable), sometimes structures our experience with programs. When we play a game, we arguably try to reverse engineer its algorithm or at the very least link its actions to its programming, which is why all design books warn against coincidence or random mapping, since it can induce paranoia in its users. That is, because an interface is programmed, most users treat coincidence as meaningful. To the user, as with the paranoid schizophrenic, there is always meaning: whether or not the user knows the meaning, s/he knows that it regards him or her. To know the code is to have a form of "X-ray vision" that makes the inside and outside coincide, and the act of revealing sources or connections becomes a critical act in and of itself.[165] Code as source leads to that bizarre linking of computers to visual culture, to transparency, which constitutes the subject of chapter 2.

Code as fetish thus underscores code as thing: code as a "dirty window pane," rather than as a window that leads us to the "source." Code as fetish emphasizes code as a set of relations, rather than as an enclosed object, and it highlights both the ambiguity and the specificity of code. Code points to, it indicates, something both specific and nebulous, both defined and undefinable. Code, again, is an abstraction that is haunted, a source that is a re-source, a source that renders the machinic—with its annoying specificities or "bugs"—ghostly. As Thomas Keenan argues, "haunting can only be thought as the difficult (simultaneous and impossible) movement of remembering and forgetting, inscribing and erasing, the singular and the different."[166] Embracing software as thing, in theory and in practice, opens us to the ways in which the fact that we cannot know software can be an enabling condition: a way for us to engage the surprises generated by a programmability that, try as it might, cannot entirely prepare us for the future.

Computers that Roar

Computers, like other media, are metaphor machines: they both depend on and perpetuate metaphors. More remarkably, though, they—through their status as "universal machines"—have become metaphors for metaphor itself.

From files to desktops, windows to spreadsheets, metaphors dominate user interfaces. In the 1990s (and even today), textbooks of human–computer interface (HCI) design described metaphors as central to "user-friendly" interfaces. Metaphors make abstract computer tasks familiar, concrete, and easy to grasp, since through them we allegedly port already existing knowledge to new tasks (for instance, experience with documents to electronic word processing). Metaphors proliferate not only in interfaces, but also in computer architecture: from memory to buses, from gates to the concept of architecture itself. Metaphors similarly structure software: viruses, UNIX daemons, monitors, back orifice attacks (in which a remote computer controls the actions of one's computer), and so on. At the contested "origin" of modern computing lies an analogy turned metaphor: John von Neumann deliberately called the major components of modern (inhuman) computers "organs," after cybernetic understandings of the human nervous system. Drawing from the work of Alan Turing and Charles Babbage, Jon Agar has argued that the computer, understood as consisting of software and hardware, is a "government machine." Like the British Civil Service, it is a "general-purpose 'machine' governed by a code."[1]

The role of metaphor, however, is not simply one way. Like metaphor itself, it moves back and forth. Computers have become metaphors for the mind, for culture, for society, for the body, affecting the ways in which we experience and conceive of "real" space: from the programmed mind running on the hard-wired brain to reprogrammable culture versus hard-wired nature, from neuronal networks to genetic programs. Paul Edwards has shown how computers as metaphors and machines were crucial to the Cold War and to the rise of cognitive psychology, an insight developed further by David Golumbia in his analysis of computationalism. As cited earlier, Joseph Weizenbaum has argued that computers have become metaphors for all "effective procedures," that is, for anything that can be solved in a prescribed number of steps, such as gene expression and clerical work.[2] Weizenbaum also notes that the power of computer as metaphor is itself based on "only the vaguest understanding of a difficult and complex scientific concept. . . . The public vaguely understands—but is nonetheless firmly convinced—that any effective procedure can, in principle, be carried out by a computer . . . it follows that a computer can at least imitate man, nature, and society in all their procedural aspects."[3] Crucially, this means that, at least in popular opinion, the computer is a machine that can imitate, and thus substitute for, all others based on its programming. This vaguest understanding—software as thing—is neither accidental to nor a contradiction of the computer as metaphor, but rather grounds its appeal.

Because computers are viewed as universal machines, they have become metaphors for metaphor itself: they embody a logic of substitution, a barely visible conceptual system that orders and disorders. Metaphor is drawn from the Greek terms *meta* (change) and *phor* (carrying): it is a transfer that transforms. Aristotle defines metaphor as consisting "in giving the thing a name that belongs to something else; the transference being either from genus to species, or from species to genus, or from species to species, or on grounds of analogy."[4] George Lakoff and Mark Johnson argue, *"The essence of metaphor is understanding and experiencing one kind of thing in terms of another."*[5] Metaphor is necessary "because so many of the concepts that are important to us are either abstract or not clearly delineated in our experience (the emotions, ideas, time, etc.), we need to get a grasp on them by means of other concepts that we understand in clearer terms (spatial orientations, objects, etc.)."[6] Lakoff and Johnson argue that we live by metaphors (such as "argument is war," "events are objects," and "happy is up"), that they serve as the basis for our thoughts and our actions.[7] Metaphors govern our actions because they are also "grounded in our constant interaction with our physical and cultural environments."[8] That is, the similarities that determine a metaphor are based on our interactions with various objects—it is therefore no accident that metaphors are thus prominent in "interactive" design. Crucially, metaphors do not simply conceptualize a preexisting reality; they also create reality.[9] Thus, they are not something we can "see beyond," but rather things necessary to seeing. Even to see beyond certain metaphors, they argue, we need others.[10] Metaphor is an "imaginative rationality": "Metaphor . . . unites reason and imagination. Reason, at the very least, involves categorization, entailment, and inference. Imagination, in one of its many aspects, involves seeing one kind of thing in terms of another kind of thing—what we have called metaphorical thought."[11] This imaginative seeing one kind of thing in terms of another thing also involves hiding: a metaphor, Thomas Keenan argues, means that "something . . . shows itself by hiding itself, by announcing itself as something else or in another form."[12]

Paul Ricoeur, focusing more on metaphor as a linguistic entity, similarly stresses the centrality and creative power of metaphor. To Ricoeur, metaphor grounds the possibility of logical thought. Ricoeur, drawing from Aristotle's definition, argues that change, movement, and transposition (and thus deviation, borrowing, and substitution) characterize metaphor.[13] By transposing an "alien" name, metaphor is a "categorical transgression . . . a kind of deviance that threatens classification itself."[14] Since metaphor, however, also "'conveys learning and knowledge through the medium of the genus,'" Ricoeur contends, "metaphor destroys an order only to invent a new one; and that the category-mistake is nothing but the complement of a logic of discovery."[15] It is a form of making, of poesis, that grounds all forms of classification.[16] This disordering that is also an ordering, a dismantling that is also a redescription, is also instructive and pleasurable—it offers us "the pleasure of understanding that follows surprise."[17] This movement from surprise to understanding is mirrored in metaphor itself, which is a mode of animation, of change—it makes things visible, alive, and actual by representing things in a state of activity.[18]

Computers, understood as universal machines, stand in for substitution itself. Allegedly making possible the transformation of anything into anything else via the medium of information, they are transference machines. They do not simply change X into Y, they also animate both terms. They create a new dynamic reality: the files they offer us are more alive; the text that appears on their screens invites manipulation, addition, animation. Rather than stable text on paper, computers offer information that is flexible, programmable, transmissible, and ever-changing. Even an image that appears stably on our screen is constantly refreshed and regenerated. Less obviously, computers—software in particular—also concretize Lakoff and Johnson's notion of metaphors as concepts that govern, that form consistent conceptual systems: software is an invisible program that governs, that makes possible certain actions. But if computers are metaphors for metaphors, they also (pleasurably) disorder, they animate the categorical archival system that grounds knowledge.

If theories of metaphor regularly assume that the vehicle (the image expressly used) makes the abstract tenor (the idea represented) concrete—that one makes something unfamiliar familiar through a known concrete vehicle—software as metaphor combines what we only vaguely understand with something equally vague. It is not simply, then, that one part of the metaphor is "hidden," but rather that both parts—tenor and vehicle—are invisibly visible. This does not mean, however, that software as metaphor fails. It is used regularly all the time because it succeeds as a way to describe an ambiguous relation between what is visible and invisible, for invisible laws as driving visible manifestations. Key to understanding the power of software—software as power—is its very ambiguous thingliness, for it grounds software's attractiveness as a way to map—to understand and conceptualize—how power operates in a world marked by complexity and ambiguity, in a world filled with things we cannot fully understand, even though these things are marked by, and driven by, rules that should be understandable, that are based on understand-ability. Software is not only necessary for representation; it is also endemic of transformations in modes of "governing" that make governing both more personal and impersonal, that enable both empowerment and surveillance, and indeed make it difficult to distinguish between the two.

2 Daemonic Interfaces, Empowering Obfuscations

Interfaces, in particular interactive GUIs (graphical user interfaces), are widely assumed to have transformed the computer from a command-based instrument of torture to a user-friendly medium of empowerment. From Douglas Engelbart's vision of a system to "augment human intellect" to Ben Shneiderman's endorsement of "direct manipulation" as a way to produce "truly pleased users," GUIs have been celebrated as enabling user freedom through (perceived) visible and personal control of the screen. This freedom, however, depends on a profound screening: an erasure of the computer's machinations and of the history of interactive operating systems as supplementing—that is, supplanting—human intelligence. It also coincides with neoliberal management techniques that have made workers both flexible and insecure, both empowered and wanting (e.g., always in need of training).[1]

Rather than condemning interfaces as a form of deception, designed to induce false consciousness, this chapter investigates the extent to which this paradoxical combination of visibility and invisibility, of rational causality and profound ignorance, grounds the computer as an attractive model for the "real" world. Interfaces have become functional analogs to ideology *and* its critique—from ideology as false consciousness to ideology as fetishistic logic, interfaces seem to concretize our relation to invisible (or barely visible) "sources" and substructures. This does not mean, however, that interfaces are simply ideological. Looking both at the use of metaphor within the early history of human–computer-interfaces and at the emergence of the computer as metaphor, it contends that real-time computer interfaces are a powerful response to, and not simply an enabler or consequence of, postmodern/neoliberal confusion. Both conceptually and thematically, these interfaces offer their users a way to map and engage an increasingly complex world allegedly driven by invisible laws of late capitalism. Most strongly, they induce the user to map constantly so that the user in turn can be mapped. They offer a simpler, more reassuring analog of power, one in which the user takes the place of the sovereign executive "source," code becomes law, and mapping produces the subject. These seemingly real-time interfaces emphasize the power of user action and promise topsight for all: they allow one to move from the local detail to the global picture—through an allegedly traceable and concrete path—by simply clicking a mouse. Conceptually drawn from auto navigation systems, these interfaces follow in the tradition of cybernetics (named after the Greek term *kybernete* for steersmen or governor) as a way to navigate or control, through a process of blackboxing.

Because of this, they render central processes for computation—processes not under the direct control of the user—daemonic: orphaned yet "supernatural" beings "between gods and men . . . ghosts of deceased persons, *esp.* deified heroes."[2] Indeed, the interface is "haunted" by processes hidden by our seemingly transparent GUIs that make us even more vulnerable online, from malicious "back doors" to mundane data gathering systems. Similar to chapter 1, this chapter thus does not argue we need to move beyond specters and the undead, but rather contends that we should make our interfaces more productively spectral—by reworking rather than simply shunning the usual modes of "user empowerment."

Interface, Intrafaith

Interactive interfaces—live screens between man and machine—stem from military projects, such as SAGE discussed in chapter 1. SAGE, according to Paul Edwards, was "a metaphor for total defense," a Cold War project that enclosed "the United States inside a radar 'fence' and an air-defense bubble."[3] Edwards describes SAGE as both based on and the basis for the world as a closed world, "an inescapably self-referential space where every thought, word, and action is ultimately directed back toward a central struggle."[4] (The opposite of a closed world is a green world, in which "the limits of law and rationality are surpassed.")[5] SAGE began as a universal cockpit simulator, but quickly evolved into a real-time network of digital computers, designed to detect incoming Soviet missiles. Unfortunately, yet not atypically, it was obsolete by the time it was completed in 1963 due to the introduction of intercontinental ballistic missiles. Despite this, SAGE is considered central to the development of computing because it fostered many new technologies, including digital real-time control systems, core-memory devices, and most importantly for this chapter, graphical user displays.

These graphical CRT interfaces were simulations of an analog technology: radar (see figure 2.1).[6] Divided into X-Y coordinates, these displays allowed the users—military personnel tracking air space—to deploy a light pen to select potential hostile aircraft tracks. This user's control of the interface and the system depended on a selective mapping that filtered as much as it represented, reducing all air traffic to blinking lines. Because of this direct real-time contact between user and computer, SAGE and the test machines associated with it are widely considered to be predecessors to personal interactive computing, albeit discontinuously (they were initially

Figure 2.1
SAGE operator at console, 1958, National Archives photo no. 342-B-003-14-K-43548

displaced by mainframes).[7] This screen, however, was an input device for the user, not for the programmers/coders, who produced taped programs that operators would load and run.

Interactive operating systems, key to making screens serve as part of an input system for all users (thus chipping away at the boundary between user and programmer), also stemmed from military funding, in particular projects related to artificial intelligence (AI). Interactivity entailed giving over to the machine tasks that humans could not accomplish. As John McCarthy, key to both AI and time-sharing operating systems (OS), explains, the LISP programming language, used in early AI projects, was designed "in such a way that working with it interactively—giving it a command, then seeing what happened, then giving it another command—was the best way to work with it."[8] Interactivity was necessary because of the limitations of procedural programming and of early neural networks. That is, by the 1960s, the naiveté behind John von Neumann's declaration that "anything that can be exhaustively and unambiguously described, anything that can be completely and unambiguously put into words, is ipso facto realizable by a suitable finite neural network" was becoming increasingly apparent.[9] Since exhaustive and unambiguous description was difficult, if not impossible, one needed to work "interactively"—not just automatically—with a computer. The alleged father of the Internet J. C. R. Licklider's vision of "Man-Computer Symbiosis" encapsulates this intertwining of interactivity and human fallibility nicely. Describing the partnership between men and computers, Licklider predicts, "man-computer symbiosis is probably not the ultimate paradigm for complex technological systems. It seems entirely possible that, in due course, electronic or chemical 'machines' will outdo the human brain in most functions we now consider exclusively within its province."[10] Similarly Jay Forrester, the force behind SAGE's development, contended, "the human mind is not adapted to interpret[] how social systems behave . . . the mental model is fuzzy . . . incomplete . . . imprecisely stated."[11] The goal, then, was to develop artificial systems to combat human frailty by usurping the human.

Given this background and the ways in which the screen screens, the emergence of user-friendly interfaces as a form of "computer liberation" seems dubious at best and obfuscatory at worst. So, why and how is it that interactive systems have become synonymous with user and machine freedom? What do we mean by freedom here? What do these systems offer and what happens when we use them?

Direct Manipulation

The notion of interfaces as empowering is driven by a dream of individual control: of direct personal manipulation of the screen, and thus, by extension, of the system it indexes or represents. Consider, for instance, the interface to Google Earth. Offering us a god's eye view, it allows us to zoom in on any location, to fly from place to place, and to even control the amount of sunshine in any satellite photo. Google Earth, however, hardly represents the world as it is, but rather a more perfectly spherical one in which it hardly ever rains (even when the Google Earth weather layer shows rain), and in which nothing ever moves, even as time goes by. Viewing these divergences from reality as failures, however, misses what makes this program so compelling: the actions it enables, the kind of dynamic mapping actions, the "top sight"—overview and zooming—it facilitates.

Google Earth, and interactive interfaces in general, follow in the tradition of "direct manipulation." According to Ben Shneiderman, who coined the term in the 1980s, "certain interactive systems generate glowing enthusiasm among users—in marked contrast with the more common reaction of grudging acceptance or outright hostility." In these systems, the users reported positive feelings, such as "mastery" over the system and "confidence" in their continuing mastery, "competence" in performing their tasks, "ease" in learning the system, "enjoyment" in using it, "eagerness" to help new users," and the "desire" to engage the more complex parts of the system. Changes in visibility and causality seem central to the creation of a truly pleased user, in particular, "visibility of the object of interest; rapid, reversible, incremental actions; and replacement of complex command language syntax by direct manipulation of the object of interest—hence the term 'direct manipulation.'"[12]

Crucially, Shneiderman posits direct manipulation as a means to overcome users' resistance: as a way to dissipate hostility and grudging acceptance and instead to foster enthusiasm by developing feelings of mastery. Direct manipulation does this by framing the problem of work from the perspective of the worker—more precisely of the neoliberal worker who decides to work—and by replacing commands with more participatory structures.[13] Direct manipulation is thus part of the "new spirit of capitalism" that the French sociologists Luc Boltanski and Eve Chiapello outline in their book of the same title. This new spirit of capitalism fosters commitment and enthusiasm—emotions not guaranteed by pay or working under duress—through management techniques that stress "versatility, job flexibility, and the ability to learn and adapt to new duties."[14] As Catherine Malabou notes, in such a system "'the leader has no need to command,' because the personnel are 'self-organized' and 'self-controlling.'"[15] In such a system, Malabou underscores, drawing from Boltanski and Chiapello, flexibility is capitulation and normative, and "everyone lives in a state of permanent anxiety about being disconnected, rejected, abandoned."[16]

Not surprisingly, the term *direct manipulation* also draws from cognitive psychology: George Lakoff and Ben Johnson use the term in relation to Jean Piaget's argument that infants "first learn about causation by realizing that they can directly manipulate objects around them."[17] That is, infants' repeated manipulations of certain objects are key to their eventual grasping of causality: that doing X will always (or usually) cause Y to happen. Relatedly, Lakoff and Johnson argue that interactions with objects also ground metaphor, since "interactional properties are prominent among the kinds of properties that count in determining sufficient family resemblance."[18] Shneiderman also offers examples of direct manipulation outside (or at least at that point outside) of computer interfaces, most importantly the steering wheel of a car:

> Driving an automobile is my favorite example of direct manipulation. The scene is directly visible through the windshield, and actions such as braking or steering have become common skills in our culture. To turn to the left, simply rotate the steering wheel to the left. The response is immediate, and the changing scene provides feedback to refine the turn. Imagine trying to turn by issuing a LEFT 30 DEGREES command and then issuing another command to check your position, but this is the operational level of many office automation tools today.[19]

Direct manipulation is thus a metaphor based on real-time analog technologies, such as a drive shaft, and their integration into a visual system. (These analog technologies, which linked steering wheel to car wheel in a mechanical cause-and-effect relation, of course are themselves being replaced by computerized drive shafts.) HCI's version of direct manipulation is never "direct," only simulated, and the mastery, as Shneiderman notes, is "felt" not possessed. This emphasis on feelings, however, reveals that the visibility of the object of interest matters less than the affective relationship established though rapid, reversible, incremental actions.

Brenda Laurel has argued this point most influentially in her classic *Computer as Theater*. According to Laurel, direct manipulation is not and has never been enough, and the strand of HCI focused on producing more and more realistic interface metaphors is wrongheaded.[20] People realize when they double-click on a folder that it is not really a folder; making a folder more "life-like" (following the laws of gravity, having it open by the user flipping over the front flap, etc.) would be more annoying than helpful. What does help, though, is direct engagement: an interface designed around plausible and clear actions. Direct engagement, Laurel contends, "shifts the focus from the representation of manipulable objects to the ideal of enabling people to engage directly in the activity of choice, whether it be manipulating symbolic tools in the performance of some instrumental tasks or wandering around the imaginary world of a computer game." This ideal engagement "emphasizes emotional as well as cognitive values. It conceives of human-computer activity as a *designed experience*"[21]— an experience designed around "activities of choice" or, more properly, making these activities feel like activities of choice.

As a designed experience, Laurel astutely insists, computer activity is artificial and should remain so.[22] That is, fabricating computer interfaces entails "creating imaginary worlds that have a special relationship to reality—worlds in which we can extend, amplify, and enrich our own capacities to think, feel, and act."[23] The computer interface thus should be based on theater rather than psychology because "psychology attempts to *describe what goes on in the real world* with all its fuzziness and loose ends, while theatre attempts to *represent something that might go on*, simplified for the purposes of logical and affective clarity. Psychology is devoted to the end of explaining human behavior, while drama attempts to represent it in a form that provides intellectual and emotional closure."[24] Importantly, Laurel's argument, even as it condemns metaphor, is itself based on metaphor, or more precisely simile: computers as theater. It displaces rhetorical substitution from the level of the interface (objects to be manipulated) to the interface as a whole; it also makes the substitution more explicit (simile, not metaphor).

Laurel's move to theater is both interesting and interested, and it resonates strongly both with Weizenbaum's parallel between programmer as lawgiver/playwright discussed previously and with Edwards's diagnosis of the computer as a metaphor of the

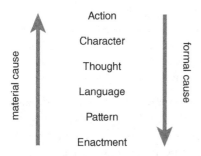

Figure 2.2
Causal relations among elements of quantitative structure. A reproduction of Brenda Laurel's illustration in *Computers as Theater*, 51.

closed world, a term also drawn from literary criticism.[25] The Aristotelian model Laurel uses provides her structuralist theory with the kind of emotional and intellectual closure she contends interfaces should create: clear definitions of causality, of the means to produce catharsis and, most important, of theater—like interfaces and computers—as following laws."[26] Clear, law-abiding causality drives every level of Laurel's system (see figure 2.2): action is the formal cause of character and so on down to enactment; enactment is the material cause of pattern and so on up to action.

Because events happen so logically, users accept them as probable and then as certain. Consequently, this system ensures that users universally suspend their disbelief. This narrowing also creates pleasure: the creation and elimination of uncertainty—the "stimulation to imagination and emotion created by carefully crafted uncertainty" and the "satisfaction provided by closure when action is complete"—Laurel contends, drives audience pleasure.[27]

The fact that users are not simply the audience, but also the actors, makes causality in computer interfaces more complicated. Thus, the designer must not simply create "good" characters that do what they intend (character, she argues, is solely defined by action), but also create intrinsic constraints so that users can become good characters too and follow the "laws" of the designer.[28] The designer is both scriptwriter and set designer: Laurel's description of the designer's power seems less extreme than Weizenbaum's; however, Laurel's vision—focused on the relationship between designer and user, rather than programmer and program—is not less but rather differently coercive. In Laurel's view, the constraints the designer produces do not restrict freedom; they ensure it. Complete freedom does not enhance creativity; it stymies it. Addressing fantasies by gamers and science fiction writers of "magical spaces where they can invent their own worlds and do whatever they wish—like gods," she argues that the experience of these spaces "might be more like an existential nightmare than a dream of freedom":

A system in which people are encouraged to do whatever they want will probably not produce pleasant experiences. When a person is asked to "be creative" with no direction or constraints whatever, the result is, according to May, often a sense of powerlessness—or even complete paralysis of the imagination. Limitations—constraints that focus creative efforts—paradoxically increase our imaginative power by reducing the number of possibilities open to us.[29]

A green world, in other words, in which action flows "between natural, urban, and other locations and centers [on] magical, natural forces" produces paralysis and nightmares. Yet constraints—the acceptance of certain interface conventions as self-enforced rules—enable agency and an arguably no less magical feeling of power: a sense that users control the action and make free and independent choices within a set of rules, again the classic neoliberal scenario. (The goal of interface design, Laurel tellingly states, is to "build a better mousetrap.")[30] To buttress this feeling of mastery, disconcerting coincidences and irrelevant actions that can expose the inner workings of programs must be eliminated. For users as for paranoid schizophrenics (my observation, not Laurel's), everything has meaning: there can be no coincidences but only causal pleasure in this closed world.

Laurel's conception of freedom, however, is disturbingly banal: the true experience of freedom may indeed be closer to an existential nightmare than to a pleasant paranoid dream. Indeed, the challenge, as I argue in *Control and Freedom: Power and Paranoia in the Age of Fiber Optics* (2006), is to take freedom seriously, rather than to reduce it to control (and thus reduce the Internet to a gated community). Freedom grounds control, not vice versa. Freedom makes control possible, necessary, and never enough. Not surprisingly, the system Laurel describes—focused on getting users to suspend disbelief and to act in certain prescribed ways—resonates widely with definitions of ideology.

Interfaces as Ideology

To elaborate on an argument I have made before, GUIs are a functional analog to ideology.[31] In a *formal* sense computers understood as comprising software and hardware are ideology machines. They fulfill almost every formal definition of ideology we have, from ideology as false consciousness (as portrayed in the 1999 Wachowski Brothers' film *The Matrix*) to Louis Althusser's definition of ideology as "a 'representation' of the imaginary relation of individuals to their real conditions of existence."[32] According to Althusser, ideology reproduces the relations of production by "'*constituting' concrete individuals as subjects.*"[33] Ideology, he stresses, has a material existence: it shapes the practices and consciousness of individual subjects. It interpellates subjects: it yells "hey you," and subjects turn around and recognize themselves in that call.

Interfaces offer us an imaginary relationship to our hardware: they do not represent transistors but rather desktops and recycling bins. Interfaces and operating systems produce "users"—one and all. Without OS there would be no access to hardware; without OS there would be no actions, no practices, and thus no user. Each OS, in its extramedial advertisements, interpellates a "user": it calls it a name, offering it a name or image with which to identify. So Mac users "think different" and identify with Martin Luther King and Albert Einstein; Linux users are open-source power geeks, drawn to the image of a fat, sated penguin (the Linux mascot); and Windows users are mainstream, functionalist types perhaps comforted, as Eben Moglen argues, by their regularly crashing computers. Importantly, the "choices" operating systems offer limit the visible and the invisible, the imaginable and the unimaginable. You are not, however, aware of software's constant constriction and interpellation (also known as its "user-friendliness"), unless you find yourself frustrated with its defaults (which are remarkably referred to as *your* preferences) or unless you use multiple operating systems or competing software packages.

Interfaces also produce users through benign interactions, from reassuring sounds that signify that a file has been saved to folder names such as "my documents," which stress personal computer ownership. Computer programs shamelessly use shifters— pronouns like "my" and "you"—that address you, and everyone else, as a subject. Interfaces make you read, offer you more relationships and ever more visuals. They provoke readings that go beyond reading letters toward the nonliterary and archaic practices of guessing, interpreting, counting, and repeating. Interfaces are based on a fetishistic logic. Users know very well that their folders and desktops are not really folders and desktops, but they treat them as if they were—by referring to them as folders and as desktops. This logic is, according to Slavoj Žižek, crucial to ideology.[34] As mentioned previously, Žižek (through Peter Sloterdijk) argues that ideology persists in one's actions rather than in one's beliefs: people know very well what they are doing, but they still do it. The illusion of ideology exists not at the level of knowledge but rather at the level of action: this illusion, maintained through the imaginary "meaning of the law" (causality), screens the fact that authority is without truth—that one obeys the law to the extent that it is incomprehensible. Is this not computation? Through the illusion of meaning and causality—the idea of a law-driven system—do we not cover over the fact that we do not and cannot fully understand or control computation? That computers increasingly design each other and that our use is—to an extent—a supplication, a blind faith?

Operating systems also create users more literally, for users are an OS construction. User logins emerged with time-sharing operating systems, such as UNIX, which encourage users to believe that the machines they are working on are their own machines (before this, computers mainly used batch processing; before that, a person really did run the computer, so there was no need for operating systems—one had human operators). As many historians have argued, the time-sharing operating systems developed in the 1970s spawned the "personal computer."[35] That is, as ideology creates subjects, interactive and seemingly real-time interfaces create users who believe they are the "source" of the computer's action.

Real-time Sourcery

According to the OED, real time is "the actual time during which a process or event occurs, especially one analyzed by a computer, in contrast to time subsequent to it when computer processing may be done, a recording replayed, or the like." Crucially, hard and soft real-time systems are subject to a "real-time constraint." That is, they need to respond, in a forced duration, to actions predefined as events. The measure of real time, in computer systems, is their reaction to the live; it is their liveness—their quick acknowledgment of and response to our actions.

The notion of real time always points elsewhere—to "real-world" events, to user's actions—thereby introducing indexicality to this supposedly nonindexical medium. That is, whether or not digital images are supposed to be "real," real time posits the existence of a source—coded or not—that renders our computers transparent. Real-time operating systems create an "abstraction layer" that hides the hardware details of the processor from application software; real-time images portray computers as unmediated connectivity. SAGE, for instance, linked computer-generated images to lines on a screen; unlike in the case of radar images, there was no "footprint" relation between screen and incoming signal. As RealPlayer reveals, the notion of real time is bleeding into all electronic moving images, not because all recordings are live, but because grainy moving images have become a marker of the real.[36] What is authentic or real is what transpires in real time, but real time is real not only because of this indexicality—this pointing to elsewhere—but also because of its quick reactions to users' inputs.

Dynamic changes to web pages in real time, seemingly at the bequest of users' desires or inputs, create what Tara McPherson has called "volitional mobility." Creating "Tara's phenomenology of websurfing," McPherson argues:

> When I explore the web, I follow the cursor, a tangible sign of presence implying movement. This motion structures a sense of liveness, immediacy, of the now . . . yet this is not just the same old liveness of television: this is liveness with a difference. This liveness foregrounds volition and mobility, creating a liveness on demand. Thus, unlike television which parades its presence before us, the web structures *a sense of causality* in relation to liveness, a liveness which we navigate and move through, often structuring a feeling that our own desire drives the movement. The web is about presence but an unstable presence: it's in process, in motion."[37]

This liveness, McPherson carefully notes, is more the illusion—the feel or sensation— of liveness, rather than the fact of liveness; the choice yoked to this liveness is similarly a sensation rather than the real thing (although one might ask: What is the difference between the feel of choice and choice itself? Is choice alone not a limited agency?). The real-time moving cursor and the unfolding of an unstable present through our digital (finger) manipulations make us crane our necks forward, rather than sit back on our couches, causing back and neck pain. The extent to which computers turn the most boring activities into incredibly time-consuming and even enjoyable ones is remarkable: one of the most popular computer games to date, *The Sims*, focuses on the mundane; action and adventure games reduce adventure to formulaic motion-restricted activities, yet the delights of interpreting these interfaces by interacting with them makes them pleasurable and never-ending. This volitional mobility, McPherson argues, reveals that the "hype" surrounding the Internet does have some phenomenological backing. This does not necessarily make the Internet an empowering medium, but at the very least means that it can provoke a desire for something better: true volitional mobility, true change.[38] Crucially, this fostering of a belief in true change—in the ability to change, in the direct causality between one's actions and a result—is programmed into the interface. That is, change, rather than being a radical act, is now the norm; we click, we change.

Interactive pleasure does not simply derive from a representation of user actions in a causally plausible manner; it also comes from "user amplification." Lev Manovich explains "user amplification" in terms of the Super Mario computer game: "When you tell Mario to step to the left by moving a joystick, this initiates a small delightful narrative: Mario comes across a hill; he starts climbing the hill; the hill turns to be too steep; Mario slides back onto the ground; Mario gets up, all shaking. None of these actions required anything from us; all we had to do is just to move the joystick once. The computer program amplifies our single action, expanding it into a narrative sequence."[39] This user amplification mimics the "instruction explosion," described in the previous chapter, central to higher-level programming languages (one line of high-level code corresponds to more than one line of machine code). User amplification also maps our actions to movements on the screen.

In essence, real-time interfaces map user actions to screened changes, making our machines seem transparent and rendering our screen into a map. Maps dominate interfaces, from our "desktop" to the clickable image maps on web pages, and mapping—the act of making and outlining connections—drives our actions online, from creating social maps based on Facebook friends to following links within web pages. Julian Dibbell has argued eloquently that online spaces are themselves essentially maps, that is, diagrams that we seek to inhabit.[40] Maps and mapping are also the means by which we "figure out" power and our relation to a larger social entity. Touchgraph's mapping of relationships between Facebook photos, Amazon books, and web pages, for instance, allegedly reveals the hidden interconnections driving consumption and social bonding (see figure 2.3).

The much celebrated theyrule.net, which allows users to map connections between people on company boards, exemplifies this notion of mapping as a form of ideology critique (see figure 2.4).

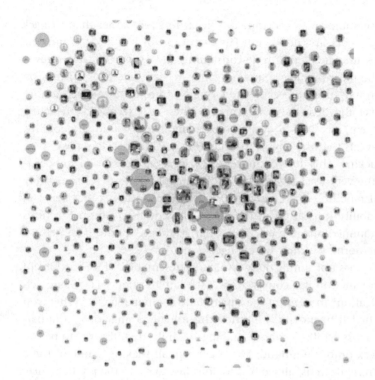

Figure 2.3
TouchGraph photos Facebook and Interactive Friends Graph, <http://blog.mememapper
.com/?p=56>, accessed 8/8/2010

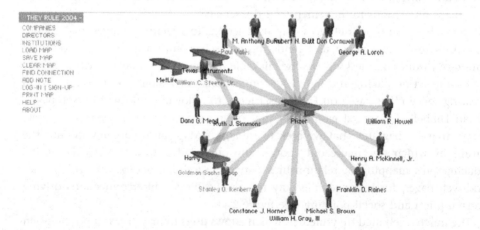

Figure 2.4
A screenshot from Theyrule.net

Indeed, Manovich argues that theyrule.net exemplifies a new rhetoric of inter-activity that "instead of presenting a packaged political message . . . gives us data and the tools to analyze it. It knows that we are intelligent enough to draw the right conclusion . . . we get convinced not by listening/watching a prepared message but by actively working with the data: reorganizing it, uncovering the connections, becoming aware of correlations." This passage intriguingly posits the program as "knowing" and the user as learning through acting. According to Manovich, this new rhetoric of interactivity is further explored in UTOPIA:

> The cosmogony of this world reflects our new understanding of our own planet—post Cold War, Internet, ecology, Gaia, and globalization. Notice the thin barely visible lines that connect the actors and the blocks. (This is the same device used in theyrule.net.) In the universe of UTOPIA, everything is interconnected, and each action of an individual actor affects the system as a whole. Intellectually, we know that this is how our Earth functions ecologically and economically—but UTOPIA represents this on a scale we can grasp perceptually.[41]

UTOPIA seemingly enables what Fredric Jameson has called a "cognitive map," a concept that I will address in more detail shortly. Briefly, it is "a situational repre-sentation on the part of the individual subject to that vaster and properly unrepre-sentable totality which is the ensemble of society's structures as a whole."[42] If cognitive mapping is both difficult and necessary now because of invisible networks of capital, these artists produce a cognitive map by exploiting the invisibility of information. The functioning of these smart interfaces parallels Marxist ideology critique. The veil of ideology is torn asunder by grasping the relations between the action of individual actors and the system as a whole. Software enables this critique by representing it at a scale—in a microworld—that we can make sense of and in which our actions and connections are amplified. This unveiling depends on our own actions, on us manipulating in order to see, on us thinking like object-oriented programmers.

It would seem thus that instead of a situation in which the production of cognitive maps is impossible, we are locked in a situation in which we produce them—or at the very least approximations of them—all the time, in which the founding gesture of ideology critique is simulated by something that also pleasurably mimics ideology. Software and ideology fit each other perfectly because both try to map the tangible effects of the intangible and to posit the intangible cause through visible cues. Both, in other words, promise a vision of the whole elephant. Through this process the invisible whole emerges as a thing, as something in its own right, and users emerge as mapping subjects.

Although the parallel between software and ideology is compelling, it is important that we not rest here, for reducing ideology to software ignores how theories of ideo-logy critique power—something essential to any theory of ideology (these resonances, however, arguably reveal the paucity of our theories of power).[43] The fact that software, with its onion-like structure (a product of programming languages), acts both as ideo-logy *and* as ideology critique—as a concealing and as a means of revealing—also breaks the analogy between software and ideology, or perhaps reveals the fact that ideology always also contains within itself ideology critique. Indeed, to take this argument further, we need to move beyond the remarkable likeness—and condemnation of screens as ideological—and to ask:[44] Under what conditions have these likenesses emerged? What, in other words, has made these likenesses and interfaces possible? What makes interfaces such a compelling imaginary map of the real? And what makes us believe that ideology is a map driven by invisible forces? Why interfaces now? And, most probingly, to what extent do interfaces stand in for likeness, for metaphor itself, and to what extent is this substitutability its most ideological aspect?

Postmodern Confusion, Interface Clarity

This drive to constantly map—and to understand through mapping—responds to postmodernist disorientation. Postmodernism, according to Fredric Jameson and Jean-François Lyotard, is/was driven by a loss of modernist certainty.[45] Lyotard defines postmodernism as an incredulity toward metanarratives (grand stories that formerly legitimated society and knowledge production). For Lyotard, this is positive because it fundamentally undermines totalitarianism and fosters creative engagement, for all actors know that legitimation—truth and justice—springs from their own creative linguistic acts. Rather than signaling the demise of existing social bonds, postmodernism promotes new social bonds since everyone (as active "nodes" in communications networks) is now involved in multiple language games. Jameson's view, however, is less optimistic. To Jameson, postmodernism or the logic of late capitalism, "is what you get when the modernization process is complete and nature is gone for good. It is a more fully human world than the older one, but one in which 'culture' has become a veritable 'second nature.'"[46] Postmodernism, Jameson contends, correlates formal changes in cultural products to a new type of social life and to a new economic order: it is "the consumption of sheer commodification as process," a transnational world in which capitalism has been completely naturalized and traditional labor placed in crisis.[47]

Postmodernism, Jameson argues, is experienced as a spatial dysfunction, as a new space that "involves the suppression of distance (in the sense of Benjamin's aura) and the relentless saturation of any remaining voids and empty places, to the point where the postmodern body . . . is now exposed to a perceptual barrage of immediacy from which all sheltering layers and intervening mediations have been removed."[48] This spatial disorientation, Jameson argues, consists of "symptoms and expressions of a new and historically original dilemma, one that involves our insertion as individual subjects into a multidimensional set of radically discontinuous realities, whose frames range from the still surviving spaces of bourgeois private life all the way to the unimaginable decentering of global capital itself." It is a new dilemma that confounds all our normal means of modeling/comprehension, making it even more difficult to understand the relation between our authentic experiences and their truth. Jameson contends, "not even Einsteinian relativity, or the multiple subjective worlds of the older modernists, is capable of giving any kind of adequate figuration to this process, which in lived experience makes itself felt by the so-called death of the subject, or, more exactly, the fragmented and schizophrenic decentering and dispersion of this last."[49] The new spaces that surround us demand that we "grow new organs . . . expand our sensorium and our body to some new, yet unimaginable, perhaps ultimately impossible, dimensions" in order to grasp our relation to totality—to make sense of the disconnect between, and possibly to reconnect, the real and the true.[50]

This decentering, this historically new dilemma, makes it impossible for us to cognitively map our relations, to realize our place in the late capitalist system.[51] Cognitive mapping combines the geographer Kevin Lynch's discussion of the ability of citizens to map the city around them with Althusser's definition of ideology. More precisely, "the conception of cognitive mapping proposed . . . involves an extrapolation of Lynch's spatial analysis to the real of social structure, that is to say, in our historical moment, to the totality of class relations on a global (or should I say multinational) scale."[52] Such a map, which Jameson in 1983 argues we did not yet have, is necessary in order to understand the totality that is capitalism; because the profit motive and the logic of capitalism set absolute barriers and limits to social changes and transformations, we need a way to comprehend its totality and our relation to it. Of anyone who does not believe that "the profit motive and the logic of capital accumulation are not the fundamental laws of this world," Jameson asserts, "such a person is living in an alternative universe."[53]

Importantly, Jameson does argue that cyberpunk and other literature/art that deals with the thematics of mechanical reproduction, as well as paranoid conspiracy theories, offer "a degraded figure of the great multinational space that remains to be cognitively mapped."[54] This is because they are figurations "of something even deeper, namely the whole world system of a present-day multinational capitalism. The technology of contemporary society is therefore mesmerizing and fascinating not so much in its own right but because it seems to offer some privileged representational shorthand for grasping a network of power and control even more difficult for our minds and imaginations to grasp: the whole new decentered global network of the third stage of capital itself."[55] A nondegraded figure, however, would be able to deal with mapping at the level of form, rather than simply content. He stresses that the call for an aesthetics of cognitive mapping

is not . . . a call for a return to some older kind of machinery, some older and more transparent national space, or some more traditional and reassuring perspectival or mimetic enclave: the new political art (if it is possible at all) will have to hold to the truth of postmodernism, that is to say, to its fundamental object—the world space of multinational capital—at the same time at which it achieves a breakthrough to some as yet unimaginable new mode of representing this last, in which we may again begin to grasp our positioning as individual and collective subjects and regain a capacity to act and struggle which is at present neutralized by our spatial as well as our social confusion.[56]

This chapter has been arguing that interfaces—with their constant emphasis on the act of making connections—*would seem* to instantiate an aesthetics of cognitive mapping. They provide a mapping—a "cognitive connectionism"—that respects the space of multinational capital and the ways in which that totality is not immediately experienceable or knowable, and yet also enables agents to act as sources. Indeed, many activists have argued that the Internet and text messaging offer effective ways of intervening on global capitalism. Rather than immobilized subjects, we have a surfeit of "produsers," who diligently produce, post, and click, providing content for "free."[57]

Interfaces in general, however, are hardly radical and the demand that we map— and thus understand—often seems like the simple following of the network and its paranoid logic rather than an insightful, clarifying act. Mapping often seduces us into exposing what is "secret" or opaque, into drawing connections between visible effects and invisible causes, rather than actually reading what one sees. It can become an endless pursuit of things, aimed at robbing them of their thingliness, in order to create a closed world in which every connection is exposed, every object reduced to a code. Interfaces are not the cognitive maps called for by Jameson because they do not engage the totality of class relations, but rather focus on totality differently figured (information networks, etc.). Whether or not interfaces are really the cognitive maps Jameson envisioned, however, is not the point here, for I do not simply want to condemn interfaces as false consciousness/false maps, but rather to understand how the once radical demand for cognitive mapping has become incorporated into the system of global capitalism/neoliberalism. As David Harvey notes, neoliberalism "requires technologies of information creation and capacities to accumulate, store, transfer, analyse, and use massive databases to guide decisions in the global marketplace."[58] The incredible proliferation of personal mapping interfaces coincides with neoliberalism's spread: these interfaces buttress notions of personal action, freedom, and responsibility.

We are now constantly called on to map and to value mapping in order to experience power/agency. This constant mapping signifies a new/neo condition, one that both recalls the power of the subject, supposedly dispersed by postmodernism, and places the subject/user in a different position than the liberal subject with respect to the "invisible" hands of the market. Liberalism traditionally challenged sovereign power, because, in the liberal market, "it is impossible for the sovereign to have a point of view on the economic mechanism which totalizes every element and enables them to be combined artificially or voluntarily."[59] Because knowledge was impossible, each subject in a market economy was supposed to act blindly, and through his or her selfishness benefit society. In a current neoliberal state (which itself is a reaction to late capitalist chaos), however, each individual must "know thyself" and others: he or she is constantly driven to make connections and to relate his or her actions to the totality.

The question then is: how can we have a form of cognitive mapping that does not engage in nostalgia for sovereign power, for the subject (now multiplied everywhere) who knows? Also: how necessary is cognitive mapping? And to what extent is the desire to map not contrary to capitalism but rather integral to its current form, especially since it is through our mappings that we ourselves are mapped? That is, to what extent *is our historically novel position not our ignorance and powerlessness, but rather our determination and our drive to know?* Could it be that rather than resort to maps, we need to immerse ourselves in networked flows—time-based movements that both underlie and frustrate maps? To respond to these large questions, let us again return to interfaces and to the dreams of progress and freedom and the minute actions that buttress them.

As We *May* Think

Interfaces respond to a crisis of knowledge that calls into question scientific and human progress. Designed to "augment human intelligence," they are steeped in a nostalgic view of machines as transparent. Interfaces recall analog machines that worked by mapping: that is, by associating one element of a set to one or more elements of another. Analog computers, which I discuss in more detail in chapter 4, were essentially models that, unlike digital computers, did not require the transforming problems into small steps answerable by a yes or a no (analog integrators actually integrated). Our digital interfaces are an analogy to an analogy. As David Mindell argues, whenever we use a mouse or look at our screens, we are engaging in activities that precede our digital computers.[60] The canonization of Vannevar Bush's article "As We May Think" within new media studies makes clear the nostalgia for an analog future.

"As We May Think," published at the end of World War II when Bush was head of the Office of Scientific Research and Development (the U.S. government agency tasked with coordinating wartime science projects, such as the Manhattan Project), is often considered the ur-text of the Internet and dynamic personal media. As Linda C. Smith among others has demonstrated, the memex, a machine for selecting and preserving data described in "As We May Think" and discussed in more detail below, is consistently and persistently cited as the inspiration for hypertext systems.[61] If William Gibson's 1984 novel *Neuromancer*, another commonly cited "precursor" to the Internet, has disappeared from new media course syllabi as new media criticism has moved away from the embarrassingly fictional and utopian, the equally fictional and utopian—the vapory—"As We May Think" has remained, because "pioneers" such as Douglas Engelbart and Ted Nelson have consistently listed it as a direct inspiration. The fact that the memex—the machine prophesized by Bush but never built—is considered a precursor, however, should make us pause, because the memex is linked to a mechanical, analog past/future that has not and arguably *may not* come to pass.[62]

"As We May Think" argues for the relevance of science to human progress. Its thesis, according to Bush's comments on an earlier draft, is that "science and its applications are not, on the whole, evil." Indeed, he ends his article by stating:

> The applications of science have built man a well-supplied house, and are teaching him to live healthily therein. They have enabled him to throw masses of people against one another with cruel weapons. They may yet allow him truly to encompass the great record and to grow in the wisdom of race experience. He may perish in conflict before he learns to wield that record for his true good. Yet, in the application of science to the needs and desires of man, it would seem to be a singularly unfortunate stage at which to terminate the process, or to lose hope as to the outcome.[63]

The threat of termination and the loss of hope Bush discusses here do not stem from politics (curiously, Bush does not mention the mounting political pressures to dismantle the "big science" machinery he established during the war), but rather from technological deficiencies. Indeed, his secondary thesis is that gadgetry is not necessarily trivial, since it "may contribute substantially to man's mental development in the future as it has in the past."[64]

Along these lines, Bush imagines the difficulties and opportunities that will face scientists during the upcoming peace in terms of "gadgetry" designed to help the scientist access the increasingly complex scientific record. A companion to Jameson's postmodern individual is the bewildered scientist, who is incapable of making sense of—of mapping—the scientific archive:

> There is a growing mountain of research. But there is increased evidence that we are being bogged down today as specialization extends. The investigator is staggered by the findings and conclusions of thousands of other workers—conclusions which he cannot find time to grasp, much less to remember, as they appear. Yet specialization becomes increasingly necessary for progress, and the effort to bridge between disciplines is correspondingly superficial.
>
> Professionally our methods of transmitting and reviewing the results of research are generations old and by now are totally inadequate for their purpose. If the aggregate time spent in writing scholarly works and in reading them could be evaluated, the ratio between these amounts of time might well be startling.[65]

Unlike Jameson, Bush's solution is mechanical rather than political, for in this article and in subsequent commentaries upon it, this second thesis supplants the first in importance, or, rather, the second becomes necessary to proving the first.[66] Without this mechanization, the scientific archive may grow, but its value will be negated, for "a record if it is to be useful to science, must be continuously extended, it must be stored, and above all it must be consulted." However, "publication has been extended far beyond our present ability to make real use of the record. The summation of human experience is being expanded at a prodigious rate, and the means we use for threading through the consequent maze to the momentarily important item is the same as was used in the days of square-rigged ships."[67] The solution Bush envisions is mechanized access: the memex.

The memex is a desk-like "gadget" with two projectors intended to enable users to make permanent associative links between documents and to retrieve them at will. The documents were to be stored as microfilm and dropped into the machine as necessary. Documents could also be added: depressing a lever would cause contents placed at the top of the memex to be photographed into the next blank space in memex film. Although the compression offered by microfilm was important, associative indexing distinguished the memex. For Bush the prime issue was selection: the human record was not being consulted because of cumbersome indexing systems. Unlike the normal alphabetical indexing systems, the memex was to create more intuitive associative trails:

> When the user is building a trail, he names it, inserts the name in his code book, and taps it out on his keyboard. Before him are the two items to be joined, projected onto adjacent viewing positions. At the bottom of each there are a number of blank code spaces, and a pointer is set to indicate one of these on each item. The user taps a single key, and the items are permanently joined. In each code space appears the code word. Out of view, but also in the code space, is inserted a set of dots for photocell viewing; and on each item these dots by their positions designate the index number of the other item.
>
> Thereafter, at any time, when one of these items is in view, the other can be instantly recalled merely by tapping a button below the corresponding code space. Moreover, when numerous items have been thus joined together to form a trail, they can be reviewed in turn, rapidly or slowly, by deflecting a lever like that used for turning the pages of a book. It is exactly as though the physical items had been gathered together from widely separated sources and bound together to form a new book. It is more than this, for any item can be joined into numerous trails.[68]

Importantly, the code space of the memex did not render these items into abstracted, disembodied information, but rather linked them together within an invisible space of place markers. The memex, in other words, was not a file system—it was not hypertext: it was a machine that did not acknowledge or create a difference between software and hardware (i.e., software, as a set of instructions that runs the machine).

The memex was an analog, mechanical—not digital—machine. Although one could argue that this was an accident of history, since Bush mainly worked with analog machines, this objection simply begs the question by assuming no substantial difference between analog and digital machines. It also ignores Bush's continued insistence that the memex was not a digital computer. Many years after the plans for the memex were first published, Bush in his "Memex Revisited," writes, "in that essay ["As We May Think"] I proposed a machine for personal use rather than the enormous computers which serve whole companies. I suggested that it serve a man's daily thoughts directly, fitting in with his normal thought processes, rather than just do chores for him."[69] Further, in discussing the question of access, Bush insisted, "for memex we need only relatively slow access, as compared to that which the digital machines demand: a tenth of a second to bring forward any item from a vast storage will do nicely. For memex, the problem is not swift access, but selective access." Moreover, he contended "we will not expect our personal machine of the future, our memex, to do the job of the great computers" and, describing the future memex's ability to "learn" and build its own trails for its master, he argued "there are already powerful mechanical [not electronic] aids."[70] This insistence on the memex as mechanical was not simply a concession to cost, but also stemmed from an understanding of the mechanical as more intuitive, more personal, as more analog and more lasting.

The memex was analogous to then current (and now resurging) models of the human mind, which, unlike models dominant during the 1960s and 1970s, did not separate the mind (software) from the brain (hardware), or assume that memories were bits of data to be manipulated algorithmically. The memex was not to model flawlessly the human mind—nor was it to be based on the fundamental "algorithm" that drove the mind—but was instead to learn from and thus act like the mind. Describing the human mind, Bush wrote, "with one item in its grasp, it [the human mind] snaps instantly to the next that is suggested by the association of thoughts, in accordance with some intricate web of trails carried by the cells of the brain. It has other characteristics, of course; trails that are not frequently followed are prone to fade, items are not fully permanent, memory is transitory."[71] The trails carried by the cells of the brain were not information—the web of trails was the mind. Further, the memex was not only to learn from but also to improve the mind: "it should be possible to beat the mind decisively in regard to the permanence and clarity of the items resurrected from storage"—the memex's traces were not to fade, making it even better than the mind. Through its permanence, it was to make an individual's "excursions" more pleasurable, since it would enable him to "reacquire the privilege of forgetting the manifold things he does not need to have immediately at hand, with some assurance that he can find them again if they prove important."[72] This permanence of the record—of microfilm—would not only grant once more the privilege of forgetting (as though any of us could ever be exempt from such a deprivation), it also would do so while saving us from repetition: repetitive thought and repetitions in thought.

According to Bush, "man" should not be burdened with repetitive thought processes like arithmetic, for which there are powerful mechanical aids. The creative aspect of thought, Bush writes, "is concerned only with the selection of the data and the process to be employed and the manipulation thereafter is repetitive in nature and hence a fit matter to be relegated to the machine." The memex could also prevent repetitive discoveries, for the danger of nonmechanically induced forgetting is repetition. In "Memex Revisited," which is itself an interesting repetition of "As We May Think," Bush contends, "an Austrian monk, Gregor Mendel, published a paper in 1865 which stated the essential bases of the modern theory of heredity. Thirty years later the paper was read by men who could understand and extend it. But for thirty years Mendel's work was lost because of the crudity with which information is transmitted between men." What is crucial—he repeats almost verbatim from "As We May Think"— is that "his [man's] situation is not improving. The summation of human experience is being expanded at a prodigious rate, and the means we use for threading through the consequent maze to the momentarily important items are almost the same as in the days of square-rigged ships." This lack of technological improvement means that "we are being buried in our own product. Tons of printed material are dumped out every week. In this are thoughts, certainly not often as great as Mendel's, but important to our progress. Many of them become lost; many others are repeated over and over and over."[73] Thus, the scientific archive rather than leading us to the future is trapping us in the past, making us repeat the present—and Bush repeat this argument—over and over again. Our product is burying us, and the dream of linear additive progress is limiting what we *may* think; but the phrase *as we may think* is richly ambiguous. At one level, it refers to a technologically enhanced future: what we *may* think if we develop prosthetic machines to supplement and access the human record, or what we may think without these devices. The word *may*, however, also refers to an authoritarian sanction—one is given the right to think X, one may think X, in which case the authority would be the machines themselves, our supposedly loyal servants. Most importantly for this argument, *may* is an uncertain link to the future: one may think this, but one is not sure. Reading against the grain of Bush's argument, I contend that this uncertainty stems not from the lack of devices such as the memex, but from the act of reading itself.

    In Bush's writing, and in prognoses of the information revolution more generally, there is no difference between access to and understanding of the record, between what would be called, perhaps symptomatically, "machine reading" and human reading and comprehension, between information and argument, between mapping and understanding. The difficulty supposedly lies in selecting the data, not in reading it, for it is assumed that reading is a trivial act, a simple comprehension of the record's content. Once the proper record is selected and the proper map produced, there is no misreading, no misunderstanding, only transparent information. If the scientific record has not been advanced, if thought is repeated, it is because something has not been adequately disseminated. Bush's argument assumes that human records make possible the construction of an over-arching archive of human knowledge in which there is no gap, no absence: a summation of human knowledge. The scientific archive thus restores, or should restore, to "mankind" everything that has eluded it.[74] So, if there is discontinuity in history, it is due to a historical accident, to our inability to adequately consult the human record, to human fallibility. This accident, however, can be solved by machines, which are viewed here as surprisingly accident-free and permanent.

A machine alone, however, cannot turn "an *information explosion* into a *knowledge explosion*";[75] it cannot fulfill the promise of what Michel Foucault has called "traditional history." Even media as stable as microfilm fade and break, and this "forgetting" of the physics of the storage medium—this conversion of medium into storage—grounds Bush's progressivist and idealist ideology. Also, as the case of Mendel reveals, the problem is not access, but rather larger epistemological frameworks. All three researchers who performed similar experiments to Mendel's thirty-five years after him consulted the scientific record and "found" Mendel, which means that Mendel's paper was not lost. The question is not why was Mendel forgotten, but rather, why, in 1900, was he remembered (and exactly what was remembered) three times independently? And why, in the history of science, is Mendel constantly being rediscovered? As Jann Sapp argues in "The Nine Lives of Gregor Mendel," this constant reinvocation is linked to the desire, on the part of reformers, to pin Mendel down as the source of *their* genetics.[76]

The example of Mendel as source is also revealing because this belief in sources— Mendel as the source of genetics, memex as the source of the Internet, code as the source of our computers—ultimately is based on a conflation of storage with access, of memory with storage, of word with action. It reduces future progress to the search for past origins. This belief also depends on our machines as being more stable and permanent, and thus better record holders, than human memory; it depends on an analogy between digital and analog media. This belief is remarkably at odds with the material transience of discrete information and the Internet.

Repetition, however, is not simply a sign of thought wasted, but also of thought disseminated. Repetition, as Derrida has argued, both makes possible and impossible the archival process: it both makes archives possible (what is contained is always an iterable representation) and, as a marker of forgetfulness, it threatens to destroy them.[77] The repetitions of Bush's goals—their adoption as forerunners of what they did not conceive—are important to understanding the emergence of interfaces as devices that empower us by reducing the world's complexity and by allowing us to forget profoundly. Computers "liberate" us from memory through their undead memories, and their interfaces mimic the workings of simple analog systems in which there is some actual connection between what we see and do, between different systems being modeled. (The analog, in this sense, mimics the repeated "citations" in Bush's texts—it links two situations.) It is not an accident that Douglas Engelbart, inventor of the mouse and widely considered to be a visionary in the development of the computer as a media machine, was not only heavily inspired by Bush's article, whose argument he arguably repeats, but also by his experiences with radar technology during World War II. In this sense, we are right to call the "real world" that our computers approximate analog, for our digital computers approximate analog computers, not only in terms of storage, but also in terms of a direct link between one's actions and the machine's, between the machine's visuals and its function. Through our originally analog mice, which translate our movements to the screen, we navigate in what seems to be "real time."

Repeating Bush

Douglas Engelbart, one of the pioneers of dynamic interactive user interfaces, known particularly for holding the patent on the mouse and for a 1968 demo now referenced as simply "the demo" or "the mother of all demos" (since it allegedly changed the lives of many who saw or even just heard of it), draws heavily from Bush for inspiration and legitimation. Indeed, Engelbart, in a letter to Bush (seeking permission to cite long sections of "As We May Think"), confessed, "I re-discovered your article about three years ago, and was rather startled to realize how much I had aligned my sights along the vector you had described. I wouldn't be surprised at all if the reading of this article sixteen and a half years ago hadn't had a real influence upon the course of my thoughts and actions."[78] Engelbart's confession reveals the extent to which technological influences are rhetorical or "vapory." In fact, although most historians and theorists focus on the content of Engelbart's work (comparing his early work to later developments), the rhetorical devices used in his texts and the semiotics of his demo are crucial to understanding the seductiveness of his vision of interactive interfaces, a vision that many derided as insane and that took many years to come into fruition.

Drawing from Bush, Engelbart developed a conceptual framework to "augment human intellect" in 1962. He first desired to create such a framework, Engelbart later explains, when he was doing odd-job electrical engineering work at Ames Research Laboratory in Mountain View, California. He was at that point several years out of school (where he studied electrical engineering) and had also had two years' experience as an electronics technician during World War II (he read Bush's "As We May Think" while stationed as a Navy boy working with radar in the Philippines). Trying to figure out what to do with his life, Engelbart recalls he had three "flashes" of insight: flash 1 was that "the difficulty of mankind's problems was increasing at a greater rate than our ability to cope. (We are in trouble.)" Flash 2 regarded his possible role in alleviating the complexity identified in flash 1: "boosting mankind's ability to deal with complex, urgent problems would be an attractive candidate as an arena in which a young person might try to 'make the most difference.' (Yes, but there's that question: of what does the young electrical engineer do about it? Retread for role as educator, research psychologist, legislator, . . . ? Is there any handle there that an electrical engineer could . . . ?)." Flash 2, therefore, focused on the question of "human capital." Engelbart's flash 3 answered the question of what a young electrical engineer could do:

> FLASH-3: Ahah—graphic vision surges forth of me sitting at a large CRT console, working in ways that are rapidly evolving in front of my eyes (beginning from memories of the radar-screen consoles I used to service).
>
>   Well, the imagery of FLASH-3 evolved within a few days to include mixed text and graphic portrayals on the CRT, and on to extensions of the symbology and methodology that we humans could employ to do our heavy thinking; and also, images of other people at consoles attached to the same computer complex, simultaneously working in a collaboration mode that would be much closer and more effective than we had ever been able to accomplish.[79]

These flashes, overwhelming pulses of light that can cause blindness, are appropriately about technological vision and images and their centrality to "governing" and improving human society. According to Engelbart, his plans to use computers as symbolic machines first met with little enthusiasm, even after he left academia to join the Stanford Research Institute (SRI). Engelbart's vision started becoming reality in 1962 when he formalized it in the SRI report "Augmenting Human Intellect: A Conceptual Framework," and in 1963 when Licklider, who had just published his article on man-machine symbiosis, provided support for Engelbart's project (while also insisting that Engelbart's system connect remotely to other computers).

   At the heart of this system of augmentation is a theory of practice, of training. According to Engelbart, we are already augmented through our use of language, customs, and tools (symbols and processes). The system, he states, "we want to improve can thus be visualized as a trained human being together with his artifacts, language, and methodology."[80] Separating technological systems from human systems, Engelbart's system seeks to produce tools to increase "the capability of a man to approach a complex problem situation, to gain comprehension to suit his particular needs, and to derive solutions to problems." This augmented system, importantly, was not simply a set of isolated tools, but "a way of life in an integrated domain where hunches, cut-and-try, intangibles, and the human 'feel for the situation' usefully co-exist with powerful concepts, streamlined terminology and notation, sophisticated methods, and high-powered electronic aids."[81] This system was thus to augment the human by producing *more* human cut-and-try technologies. Training, Engelbart stresses, superficially divides human cultures. He states: "while an untrained aborigine cannot drive a car through traffic, because he cannot leap the gap between his cultural background and the kind of world that contains cars and traffic, it is possible to move step by step through an organized training program that will enable him to drive effectively and safely. In other words, the human mind neither learns nor acts by large leaps, but by steps organized or structured so that each one depends upon previous steps."[82] At the base of Engelbart's system is a trainable exemplary "primitive" who can, through step-by-step (digital?) training, improve his or her skills.

This example of navigating a car—this comparison between digital and analog navigational systems—that was repeated by Shneiderman later is not accidental, but rather central to the conceptualization of individual interfaces. Analog technology is also embedded in what is considered to be Engelbart's most important contribution: the mouse. The mouse is based on the integraph (further discussed in chapter 4), an "analog" device designed to integrate distance based on speed. Engelbart tied his system conceptually to automobiles and to their transformation of mass-transportation systems into mass-individual systems:

> I suggest that the parallel of the individually manned auto-motive vehicles will develop in the computer field, contributing to changes to our social structure that we can't comprehend easily. The man-machine interface that most people talk about is the equivalent of the locomotive-cab controls (giving a man better means to contribute to the big system's mission), but I want to see more thought on the equivalent of the bulldozer's cab (giving the man maximum facility for directing all that power to his individual task).[83]

Rather than a system designed to move masses en masse, these interfaces personalize mass movement and destruction. It is everyone and all in a bulldozer; everyone and all's actions amplified. Engelbart's system underscores the key neoliberal quality of personal empowerment—the individual's ability to see, steer, and creatively destroy—as vital to societal development. Not surprisingly, he views his augmented lifestyle as replacing our "clerks" or personal "slaves" with computers.[84]

The stress on the individual and individual understanding is underscored in an intriguing "hypothetical description" section included in his report. Written "to give you (the reader) a specific sort of feel for our thesis," it describes "what might happen if you were being given a personal discussion-demonstration by a friendly fellow (named Joe) who is a trained and experienced user of such an augmentation system." This section, in other words, constantly interpellates "you" as a potential user of the system. Starting with a description of Joe at his workstation, it narrates not just your actions, but also your emotions. It scripts your involvement and pleasure in the learning process. For instance, Joe says, "Let's actually work some examples. You help me" after which "you become involved in a truly fascinating game." This fascinating game, in which you "help" your teacher, asks "you" to summarize what you've learned thus far about augmentation using augmentation (with a little coaching from Joe). As "you begin self-consciously to mumble some inane statements about what you have seen, what they imply, what your doubts and reservations are, etc." Joe "mercilessly ignores your obvious discomfort and gives you no cue to stop, until he drops his hands to his lap after he has filled five frames with these statements (the surplus filled frames disappeared to somewhere—you assume Joe knows where they went and how to get them back)." Through this procedure, Joe reveals "how you wandered down different short paths, and crisscrossed yourself a few times" because "you haven't been making use of the simple symbol-manipulation means that I showed you—other than the shorthand for getting the stuff on the screens." Joe then goes on to show you effective tricks that are deliberately not "impressive," for Joe's point is to make you realize that new tricks are all based on lots of changes to the little things you do. He gets you to edit, reword, compile, and delete, at which point:

> You are quite elated by this freedom to juggle the record of your thoughts, and by the way this freedom allows you to work them into shape. You reflected that this flexible cut-and-try process really did appear to match the way you seemed to develop your thoughts. Golly, you could be writing math expressions, ad copy, or a poem, with the same type of benefit. You were ready to tell Joe that now you saw what he had been trying to tell you about matching symbol structuring to concept structuring—when he moved on to show you a succession of other techniques that made you realize you hadn't yet gotten the full significance of his pitch.[85]

This feeling of freedom "you" experience stems from an increase in productivity made possible by the match, or analog, between the machine's processes and your own; this match is a "benefit" that could improve all your activities, from work to play. In typical neoliberal fashion, this report evaluates all activities in terms of a cost–benefit analysis.[86] Also, the system user is convinced not simply through doing (even hypothetical doing), but rather through interpersonal interactions, in which "you" are relentlessly coached and cajoled. Key to "human augmentation" is the establishment of users who act and through their actions believe—all via a linear narrative that praises nonlinear processes as empowering.

Figure 2.5
A screenshot from Douglas Engelbart, "The Demo," <http://video.google.com/videoplay?do cid=-8734787622017763097>, accessed 8/8/2010

Engelbart's demo, the famous "mother of all demos," similarly interpellates the viewer, using the linear conventions of live TV and cinema. Engelbart and his crew, some of whom spoke live from their Silicon Valley location, appeared on a massive twenty-two-by-eighteen-foot screen (see figure 2.5). Indeed, Engelbart begins by apologizing to his audience for addressing them mainly through the screen, and throughout the demo his face and hands and those of his colleagues fade in and out with the contents of their screen (see figures 2.6 and 2.7).

The directness of Engelbart's address, however, compensates for the screen. He describes his research project with the question: "If, in your office, you, as an intellectual worker, were supplied with a computer display backed up by a computer that was alive for you all day and was instantly [sic] responsible . . . responsive to every action that you had, how much value could you derive from that?" To answer this question, he shows you how he begins his workday, with a blank screen (which becomes your monitor/interface). He then starts to input words into a file to show you the various "view control" features of the NLS (online system); in this scenario, Engelbart becomes both Joe and you the viewer. Using the mouse (its first public display), he copies and moves texts, gradually ordering a grocery list allegedly created in response to a call from his wife; he also offers a map of his journey home. He shows how the system logs ownership and changes to the files (importantly, the NLS system also permanently saved all files). Switching to his colleagues in Silicon Valley, he has them explain more features of the system and, in one comic moment, loses the ability to speak to them, but they nonetheless continue their explanations.

Engelbart serves as our protagonist, with whom we as "intellectual workers" are supposed to identify. The view of his hands, for instance, makes our and his gaze coincide. Supplementing this cinematic call to identify with Engelbart, however, is a televisual structure of technologically mediated liveness and interpersonal discourse.[87] Engelbart looks at us directly and, like a news anchor, controls the screen, determining which of his colleagues appear next. Intriguingly, the interface itself did not have windows—everything is shown on one screen—but the notion of windowing exists in the split screens and transfers. Our screen, in other words, becomes a window in which Engelbart's face appears and disappears: a medium in the medium. He addresses us directly, making us part of his world, through his interface and through the NLS system, which magically bring us all together. The demo, in other words, has been so "life changing" not simply because of the technology it featured, but also because of the images and visions of interconnectivity it established through its visual presentation.

Through our identification with Engelbart via his demo we emerge as sovereign subjects—subjects of files. Not accidentally, Engelbart's tasks are administrative: compiling lists, assigning ownership to files. Lists, according to Cornelia Vismann,

Figures 2.6
A screenshot from Douglas Engelbart, "The Demo"

Figure 2.7
A screenshot from Douglas Engelbart, "The Demo"

are fundamentally administrative. They are about power through writing: they "do not communicate, they control transfer operations."[88] Similarly, files, which, Vismann argues, "at their core . . . are governed by lists," are central to legal institutions and power.[89] Files "are comprehensive recording devices that register everything in the medium of writing, even that which is not writing."[90] The modern interface, by putting everyone in control of their files, makes every system user a "chancellor"—again, *an executive*—and is part of an ongoing personalization of bureaucracy: "by condensing an entire administrative office, the computer implements the basic law of bureaucracy according to which administrative techniques are transferred from the state to the individual."[91] The personalizing of files—both virtually and legally through various "access to information" laws—individualizes and totalizes.

This notion of individual yet total is also underscored in the Engelbart demo format, in the ways it differs from live television and cinema. The "view control" offered by the NLS system was mirrored intriguingly in the demo itself, with the camera offering us views that were not under Engelbart's control and featuring moments of confusion and disconnection. Engelbart falters repeatedly in the demo: in the rather telling slip of the tongue listed earlier he contends that the machine is "responsible for" rather than "responsive to" our every action. He also makes mistakes and claims that he (not the machine) has not yet "warmed up." All these errors combined with the various visual views of the team place us, the viewers, in the position of control. Like the mouse that responds to Engelbart's movements, the viewer seemingly toggles, cuts, and pastes through the various views. Engelbart is not only the subject, but also an object to be manipulated.

At the alleged origins of interactive real-time interfaces, then, is a desire to control, to "govern," based on a promise of transparent technologically mediated contact. It is a vision of permanence and flexibility: the files are permanently stored and the user's information tracked. Through this, this spectral interface has come to stand in for the machine itself, erasing the medium as it proliferates its specters, making our machines transparent producers of unreal visions—sometimes terrifying but usually banal imitations or hallucinations of elsewhere, in which the uneasy relationship between human agency and dependency is negotiated.

Daemonic Media

This spectrality makes our media daemonic: inhabited by invisible, orphaned processes that, perhaps like Socrates's *daimonion*, help us in our times of need. They make executables magic. UNIX—that operating system seemingly behind our happy spectral Mac desktops—runs daemons; daemons run our email, our web servers. Macs thus not only proudly display that symbol of the Judeo-Christian seduction and fall from grace—that sanitized but nonetheless tellingly bitten apple—it also inhabits its operating systems with daemons that make it a veritable "Paradise Lost." The mascot for FreeBSD, the robust operating system distributed via Berkeley Software Distribution and descended from AT&T's UNIX (also used by Apple OSX), nicely features the daemon as its logo (see figure 2.8).

These daemons called sendmail if not Satan are processes that run in the background without intervention by the user (usually initiated at boot time). They can run continuously, or in response to a particular event or condition (for instance, network traffic), or at a scheduled time (every five minutes or at 05:00 every day).

Figures 2.8
The FreeBSD mascot, <http://www.zer0.org/daemons/wc/standing_daemon.jpg>, accessed 8/8/
2010. Copyright 1988 by Marshall Kirk McKusick, used with permission. Reprinted with
permission.

More technically, UNIX daemons are parentless—that is, orphaned—processes that
run in the root directory. You can create a UNIX daemon by forking a child process
and then having the parent process exit, so that INIT (the program that spawns
all other programs and thus the daemons of daemons) takes over as the parent
process.[92]

UNIX daemons supposedly stem from the Greek word *daemon* meaning, according
to the OED, "a supernatural being of a nature intermediate between that of gods and
men; an inferior divinity, spirit, genius (including the souls or ghosts of deceased
persons, esp. deified heroes)." A daemon thus is already a medium, an intermediate
value albeit one that is not often seen. The most famous daemon is perhaps Socrates's
daimonion—a mystical inner voice that assisted the philosopher in times of crisis by
forbidding him to do anything rash. The other famous daemon, more directly related
to those spawning UNIX processes, is known as *Maxwell's daemon*. According to Fernando Corbato, one of the original members of the Project MAC group in 1963: "Our
use of the word daemon was inspired by Maxwell's daemon of physics and thermo-
dynamics. (My background is Physics.) Maxwell's daemon was an imaginary agent
which helped sort molecules of different speeds and worked tirelessly in the background. We fancifully began to use the word daemon to describe background processes

which worked tirelessly to perform system chores."[93] Daemonic processes, therefore, are slaves that work tirelessly and, like all slaves, define and challenge the position of the master.

The introduction of multiuser command line processing—seeming if not actual real-time operating systems—necessitates the mystification of processes that appear to operate automatically without user input, breaking the interfaces' "diegesis." What is not seen becomes daemonic, rather than what is normal, because the user is supposed to be the cause and end of any process. Interactive operating systems, such as UNIX, transform the computer from a machine run by human operators in batch mode to "alive" personal machines, which respond to the user's commands. Real-time content, stock quotes, breaking news, and streaming video similarly transform personal computers into personal media machines. These moments of "interactivity" buttress the notion of our computers as transparent. Real-time processes, in other words, make the user the "source" of the action, but only by orphaning those writing processes without which there could be no user.[94] By making the interface "transparent" or "rational," one creates daemons, which as autonomous operations call into question the subject they allegedly support. It is not surprising then that Friedrich Nietzsche condemned Socrates so roundly for his daemon (and, similarly, language for its attribution of subject to verb), even though daemons are symptoms rather than causes. According to Nietzsche, Socrates was himself a daemon because he insisted on the transparency of knowledge, because he insisted that what is most beautiful is also most sensible. Crucially, Socrates's divine inner voice only spoke to dissuade. Socrates introduced order and reified conscious perception, making instinct the critic, and consciousness the creator. As a symptom of this desire for the transparency of knowledge, for the reigning of rationality, daemon is also a backronym. Since the first daemon automatically made tape backups for the file system, it has been widely and erroneously assumed that daemon initially stood for "Disk And Executive MONitor" (this alleged "source" phrase was later adopted). The first daemon appropriately is about memory: an automated process that transfers data between secondary and tertiary memory. Memory is what makes daemons possible and what makes our media daemonic. Memory, as I elaborate later, grounds code as logos.

Ghostly Interfaces, Confused Mappings

The drive to map seeks to clear up confusion and establish the user as the sovereign subject, in control of what she sees: she controls technology that transparently reveals her relationship to the invisible laws of computation. This compelling relationship bleeds elsewhere, making the interface not simply based on metaphor, but also a compelling metaphor for understanding all invisible laws. This mapping makes us believe that the world, like the computer, really comprises invisible hands and rules that we can track via their visual manifestations. Hence the popularity of software as a metaphor for almost everything—culture, genetics, life—and the reduction of everything to transparent (and in Baudrillard's term "obscene") communication. Although digital imaging certainly plays a role in the notion of computer networks as transparent, it is neither the only nor the key thing. Consider, for instance, "The Matrix" (Multistate Anti-TerRrorism Information eXchange) data mining program that sifts through databases of public and private information ostensibly to find criminals or terrorists. The Matrix works by integrating "information from disparate sources, like vehicle registrations, driver's license data, criminal history and real estate records and analyzing it for patterns of activity that could help law enforcement investigations. Promotional materials for the company that developed The Matrix put it this way: 'When enough seemingly insignificant data is analyzed against billions of data elements, the invisible becomes visible.'"[95] Although supporters claim that The Matrix simply brings together information already available to law enforcement, "opponents of the program say the ability of computer networks to combine and sift mountains of data greatly amplifies police surveillance power, putting innocent people at greater risk of being entangled in data dragnets. The problem is compounded, they say, in a world where many aspects of daily life leave online traces."[96] By March 15, 2004, over two-thirds of the states withdrew their support for The Matrix, citing budgetary and privacy concerns. The Matrix was considered to be a violation of privacy or a making of the invisible visible (again, the act of software itself) not because the computer reproduced indexical images but rather because it enabled the police to make easy connections and thus amplified their power. As mentioned previously, the Total Information Agency (TIA), formed to bring together the U.S. government's various electronic databases, was similarly decried, although not terminated (many of its programs were implemented by the NSA and continue in the Obama administration).

This desire to see and to map, however, is not limited to governmental organizations; it is also key to the increasing personalization of commercial media. In spring 2009, Cablevision announced plans to use a targeting technology, "Visible World," to "route ads to specific households based on data about income, ethnicity, gender or whether the homeowner has children or pets" to 500,000 homes in the Tri-State region surrounding greater New York and encompassing the populated areas of New York, New Jersey, and Connecticut (Cablevision has already been testing this new technology within a sample urban population over the past year and a half). Visible World" aims to make television—not the user—smarter: "Television was always big and dumb," said Seth Haberman, the chief executive of Visible World. "Now, hopefully, we can be big and slightly smarter."[97] In this definition, *smart* means tracking users' actions: like the iTunes and Amazon.com user-driven product recommendation systems, smart technologies automatically capture information about users' personal preferences and usage and send it to a central database to be processed and analyzed.[98]

To continue on a more personal level, computing as enabling connections and making the invisible visible drives personal computing interfaces. By typing in Microsoft Word, letters appear on my screen, representing what is stored invisibly on my computer. My typing and clicking seem to have corresponding actions on the screen. By opening a file, I make it visible. On all levels, then, software seems to be about making the invisible visible—about *translating* between computer-readable code and human-readable language. Manovich seizes on this translation and makes "transcoding"—the translation of files from one format to another, which he extrapolates to the relationship between cultural and computer layers—his fifth and last principle of new media in *The Language of New Media*. Manovich argues that in order to understand new media we need to engage both layers, for although the surface layer may seem like every other media, the hidden layer, computation, is where the true difference between new and old media—programmability—lies. He thus argues that we must move from media studies to software studies, and the principle of transcoding is one way to start to think about software studies.[99]

The problem with Manovich's notion of transcoding is that it focuses on static data and treats computation as a mere translation. Not only does programmability mean that images are manipulable in new ways; it also means that one's computer constantly acts in ways beyond one's control. To see software as merely transcoding erases the computation necessary for computers to run. The computer's duplicitous reading does not simply translate or transcode code into text/image/sound, or vice versa; its reading—which conflates reading and writing (for a computer, to read is to write elsewhere)—also partakes in other invisible readings. For example, when Microsoft's Media Player plays a CD, it sends the Microsoft Corporation information about that CD. When it plays a Real Media file, such as a CNN video clip, it sends CNN its "unique identifier." You can choose to work offline when playing a CD and request that your media player not transmit its "unique identifier" when online, but these choices require two changes to the default settings. By installing the Media Player, you also agreed to allow Microsoft to "provide security related updates to the OS Components that will be automatically downloaded onto your computer. These security related updates may disable your ability to copy and/or play Secure Content and use other software on your computer." Basically, Microsoft can change components of your operating system (OS) without notice or your explicit consent. Thus to create a more "secure" computer, where secure means secure *from* the user, Microsoft can disable pirated files and applications and/or report their presence to its main database.[100] Of course, Microsoft's advertisements do not emphasize the Media Player's tracking mechanisms but rather sell it as empowering and user friendly. So now you can listen to both your CD and Internet-based radio stations with one click of a mouse: it is just like your boom box, but better. And now you can automatically receive software updates and optimize your connection to remote sites. As mentioned previously, this logic also drives Apple's iTunes recommendation system; more troublingly, it also appears in Apple's DRM-free music tracks, which embeds personal account information into these freely accessible tracks.

To be clear: this chapter is not a call to a return to an age when one could see and comprehend the actions of our computers. Those days are long gone. As Friedrich Kittler argues, at a fundamental level we no longer write; through our use of word processors we have given computers that task.[101] Neither is this chapter an indictment of software or programming (I too am swayed by and enamored of the causal pleasure of software). It is, however, an argument against common-sense notions of software precisely because of their status as common sense (and in this sense they fulfill Antonio Gramsci's notion of ideology as hegemonic common sense); because of the histories and gazes such notions erase; and because of the future they point toward. Software has become a common-sense shorthand for culture, and hardware shorthand for nature. (In the current debate over stem cell research, stem cells have been called "hardware." Historically software also facilitated the separation of pattern from matter, necessary for the separation of genes from DNA.)[102] In our so-called post-ideological society, software sustains and depoliticizes notions of ideology and ideology critique. People may deny ideology, but they don't deny software—and they attribute to software, metaphorically, greater powers than have been attributed to ideology. Our interactions with software have disciplined us, created certain expectations about cause and effect, offered us pleasure and power—a way to navigate our neoliberal world—that we believe should be transferable elsewhere. It has also fostered our belief in the world as neoliberal: as an economic game that follows certain rules. The notion of software has crept into our critical vocabulary in mostly uninterrogated ways.[103] By interrogating software and the visual knowledge it perpetuates, we can move beyond the so-called crisis in indexicality toward understanding the new ways in which visual knowledge—seeing/visible reading as knowing—is being transformed and perpetuated, not simply rendered obsolete or displaced.

To do this, we need, again, to understand the ways in which the drive to map and to promote transparency enables nontransparent data tracking that cuts across the governmental, the political, the commercial, and the personal. For instance, Google's 2009 decision to refuse to self-censor its search results in China seems a simple rejection of governmental transcoding and a laudable endorsement of transparency: by refusing to censor results, Google makes visible what is usually invisible to or screened from its Chinese users. Google made its decision, however, in response to the fact that "Chinese hackers had penetrated some of its services, such as Gmail, in a politically motivated attempt at intelligence gathering."[104] Importantly, Google's complicity with the U.S. government's email surveillance program made possible this spying foray: the hackers took advantage of the fact that, "in order to comply with government search warrants on user data, Google created a backdoor access system into Gmail accounts."[105] Similarly, the U.S. government's celebration and endorsement of social networking tools such as Twitter makes Twitter users open to governmental surveillance: by using Twitter and other U.S. software, one makes it much easier for the U.S. government to track one's actions through "back doors" (back doors created in part through our belief in our interfaces as simple windows).

This does not mean, however, that one must simply condemn all attempts at political transparency and mapping and all user actions as self-delusional capitulation. In response to TIA, Chris Csikszentmihalyi's group at the MIT Media Lab produced a remarkable site, Government Information Awareness (GIA), which mobilized resistance against TIA by reversing the gaze. Rather than facilitating information gathering about citizens, GIA sought to create a sort of "citizen's intelligence agency" that gathered data on elected officials, empowering "citizens by providing a single, comprehensive, easy-to-use repository of information on individuals, organizations, and corporations related to the government of the United States of America."[106] Although the GIA is no longer functional, its drive to make information public regarding governmental actions has been embraced by the Obama administration itself. This administration—even as it supported "back doors" in email and social networking software—created the Open Government Initiative, which is committed to making government transparent, participatory, and collaborative. Csikszentmihalyi's group has further pursued these issues of participation and transparency through numerous projects, such as Xtract, which give citizens access to data that is usually limited to commercial and governmental organizations and, most important, gives them the ability to manipulate and update these data. These projects are critical not simply for the "transparency" or information they seem to offer, but also because they give users the collective power to transform these databases and these debates over what counts as evidence.

This interrogation of interfaces also entails addressing how their mapping never works as simply as projected: there is a reason why demos—with their directed narrative and limited interactivity—are so compelling and so central to the computer industry. Indeed, the action of clicking, and that of "driving," from one site to another, as I have argued elsewhere, is more like an itinerary—a journey without a global view—than a map. That is, if our interfaces are maps, we don't necessarily treat them as such: we try to inhabit them, and, by inhabiting them, we turn them into something other than a map. Significantly, though, these journeys, or tracings, can always be reincorporated into a map, and every journey we take, through the storage and sites we use, can be recompiled in a map that allegedly contains the truth of our journey—hence Google's ability to track flu outbreaks based on search terms entered into its database, although again, databases are not as infallible as they seem.[107] These databases, which drive computer "mapping" / machine intelligence, become "dirty," unreliable, when they do not actively erase information: they become flooded with old and erroneous information that dilutes the maps they produce. Deliberately making databases dirty—by providing too much or erroneous information—may be the most effective way of preserving something like privacy.

Furthermore, the more an interface is programmed—the more it tries to meet and anticipate users' needs and to direct their actions—the more confused and confusing it becomes. As Matthew Fuller has argued in relation to Microsoft Word's massive feature mountain (opening all the possible menus and tools, for instance, completely blocks the screen so we cannot see what we are writing), the more features that are offered, the more the user can go astray. This going astray is not necessarily a bad thing—it does not necessarily lead to an existential nightmare—or, if it does, perhaps that is a good thing. What certain types of mapping and actions eradicate is the ways in which we are not the only agents: as Adrian Mackenzie argues, software is a neighborhood, an amalgam that brings together many different modes of action.[108] As a neighborhood, software entails all those "neighborhood effects"—effects by individuals on others for which "it is not feasible to charge or recompensate them"—that even neoliberals acknowledge belie freedom as a strictly voluntary exchange.[109]

In addition—as Csikszentmihalyi's projects make clear—metaphor and mapping are creative acts. That is, rather than simply condemning maps that seek to reduce the earth to a spinning globe, we need to embrace the ways in which mapping and metaphor—as artificial acts—create the world they represent. Friedrich Nietzsche, for instance, has argued that all language—and truth—is basically metaphorical. What is truth, he asks, if not

> a mobile army of metaphors, metonyms, and anthropomorphisms—in short, a sum of human relations which have been enhanced, transposed, and embellished poetically and rhetorically, and which after long use seem firm, canonical, and obligatory to a people: truths are illusions about which one has forgotten that this is what they are; metaphors which are worn out and without sensuous power; coins which have lost their pictures and now matter only as metal, no longer as coins.[110]

Rather than forgetting this "primitive world of metaphor" and thus live securely and consistently, Nietzsche calls for an embrace of the fundamental drive to form metaphors and thus to refashion the world. Understanding computers as metaphors for metaphor also means engaging the artificiality of metaphor—producing new metaphors—that make strange and estranging the world around us. This aesthetic creation would not seek to make a totality visible, so we can then navigate it, but rather, through deliberately odd and artificial means, to create what we consider totality to be. This, arguably, is the value of the many aesthetic "detournements" of mapping produced by artists, from i/o/d's *WebStalker* (which offers its "surfers" HTML code and a map of a web pages rather than its "content") to Jonah Brucker-Cohen's *Wireless*

Hog (which offers users a way to interfere with private wireless networks). Liz Canner and John Ewing's 2001 *Symphony of a City* nicely demonstrates the possibilities of this mapping. Canner and Ewing's project addressed the housing crisis in Boston by attaching portable cameras to the heads of eight nominated "representatives" of differently affected communities (including a housing court lawyer, slum lord, homeless person, and an artist). Canner and Ewing then streamed the camera views live on the Web and projected them (as four quadrants) simultaneously on the side of Boston City Hall. This move to reclaim a "monument" within Boston—a city crucial to Lynch's analysis—offers a provocative remapping of city space that links individual experience to the larger issues of gentrification and capital. It also allows one to "follow" this issue by literally following the flow of the participants: there is no god's eye view, but rather a splitting of perspectives that demands both local attention and global distraction.

Last, this interrogation of transparency, mapping, and interfaces needs to address the ways in which the digital rather than simply offering a stable material for memory is also fundamentally ephemeral. The digital, if it is anything, is the enduring ephemeral. Digital media is not always there (accessible), even when it is (somewhere). We suffer daily frustrations with digital "sources" that just disappear. Digital media is degenerative, forgetful, erasable. This degeneration makes it both possible and impossible for it to imitate analog media, making it perhaps a device for history, but only through its ahistorical (or memoryless) functioning, through the ways in which it constantly transmits and regenerates text and images. The question becomes: how did constant regeneration become stable transmission?

II Regenerating Archives

The transience of new archives, their ever shorter half-life, is their fate, their curse, and their opportunity.
—Wolfgang Ernst[1]

Traces . . . produce the space of their inscription only by acceding to the period of their erasure.
—Jacques Derrida[2]

A major—if not the major—category of new media is memory.[3] Memory, a metaphor become essence, is assumed to be its ontology at all levels, from hardware to software, from content to purpose. From CD ROMs (compact disc read-only memory) to memory sticks, from RAM (random-access memory) to ROM (read-only memory), computer hardware is riven by memory, which, as I elaborate later, makes porous the boundaries of the machine. Memory underlies the emergence of the computer as we now know it: the move from calculator to computer depended on "regenerative memory."[4] The Internet's content, memorable or not, is similarly shot through with memory. Many web sites and digital media projects focus on preservation: from online museums to YouTube phenomenon Geriatic1927, from Bill Gates's Corbis Image Database to the Google databanks that store every search ever entered (and link them to an IP address, allegedly making Google the "stasi resource" of the twenty-first century). Memory hardens information—turning it from a measure of possibility into a "thing," while also erasing the difference between instruction and data (computer memory treats them indistinguishably). It seems to make digital media an ever-increasing archive in which no piece of data is lost and thus central to progress.

This always thereness of new media links it to the future as future simple. By saving the past, it is supposed to make knowing the future easier. The future is technology because technology enables us to see trends and hence to make projections—it allows us to intervene on the future based on stored programs and data that compress time and space.[5] Again, investment banks, such as Goldman Sachs, pay millions just to be one millisecond closer to the future than their competitors. More damningly, new media allegedly puts the future simple into place through the threat of constant exposure. As a *New York Times* article questioned in response to the seminal posting of the Senator George Allen *macaca* clip to YouTube in 2006: "If . . . any moment of a candidate's life can be captured on film and posted on the Web, will the last shreds of authenticity be stripped from our public officials?"[6] Intriguingly, this formulation assumes that racist slurs are the authentic and the true, and that public exposure always makes behavior more banal. However, given the legion of students with compromising Facebook entries who seem oblivious to the fact that potential employers can check these entries, and given that people increasingly record their own "transgressions" (such as the English happy slappers who taped themselves accosting unsuspecting people in buses etc. and posted them to YouTube), it is not clear that this assumption will hold, even for politicians. Allen, after all, directed his comment at a public rally to an Indian-American man holding a video recorder. Regardless, digital media was supposed—in its very functioning—to encapsulate the enlightenment ideal that better information leads to better knowledge, which in turn guarantees better decisions.[7] As a product of programming, digital media was to program the future.

The next two chapters flesh out the emergence of computer memory and its importance to notions of programmability. Focusing on the relation between biology and computing technology, chapters 3 and 4 explore how something so initially "vapory" as software came to embody the logic of the "always already there." By exploring the ways in which biology and computer technology have been reduced to complementary strands of a double helix, they embed computer technology within the larger epistemic field of programmability. Both are a return to a reductionist, mechanistic understanding of life, in which the human body becomes an archive. Both are the basis of a biopolitics that seeks to rationalize and optimize human populations and capital.

Chapters 3 and 4 also explore how permanence has become intertwined with undead repetition and transmission, and how this repetition both grounds and threatens computers as archive. Repetition, as Vannevar Bush and Jacques Derrida have argued, is also a marker of forgetfulness. Drawing from Freud's work on the death drive, Derrida contends "repetition itself, the logic of repetition, indeed the repetition compulsion, remains . . . indissociable from . . . destruction."[8] This repetition also relies on degeneration: as the epigraph states, a trace to be a trace must be erasable. Because of this, it is also undead or spectral, neither human nor machine, "neither present nor absent 'in the flesh,' neither visible nor invisible, a trace always referring to another whose eyes can never be met."[9] It is, to tie it to the first section, invisibly visible, visibly invisible, a "technical" repetition that relates life and death, presence and representation.[10]

By focusing on archives—on the relation between memory and repetition, repetition and forgetting, repetition and transmission—this section also addresses questions of power. Vannevar Bush's insistence that accessing and wielding the archive is key to the survival of the human race may seem hyperbolic, but archives have historically been linked to questions of authority. As Howard Caygill, paraphrasing Aristotle, explains:

> The institution of the *archon* originated in the ancient Greek transition from monarchic to aristocratic rule, with the *archons*, unlike the kings, being constitutionally required to respect precedent. On assuming office an *archon* had to make a proclamation 'that whatever each man possessed before his entry into office he shall possess and control until the end of it. . . .' In order to honour such a commitment it was necessary to preserve authoritative records of 'what each possessed' at the beginning of each archonate, and these records, or rather the building in which they were stored, became known as the archive. It is important to recall the origins of the archive in oligarchic rule, because it is characteristic of such regimes that the laws be public, but not available to all.[11]

The archive thus buttresses a certain definition of public as state authority through the transformation, as Derrida notes, of a private domicile into a public one. It is also based on a promise that links the past to the future: whatever is possessed at the beginning of an *archon*'s term shall remain at the end; an archive conserves. This conservative promise is tied to another: the promise to respect precedent, that is, to follow past rules in order to guarantee a just future. Derrida thus argues, "the archive is a pledge to the future, it is not an issue of the past: it is a question of the future, the question of the future itself, the question of a response, of a promise, and of a responsibility for tomorrow. The archive: if we want to know what that will have meant, we will only know it in times to come."[12] The meaning of an archive, like source code, can only be determined after the fact. It is a promise to the future.[13] Derrida also argues, "there is no political power without control of the archive, if not of memory. Effective democratization can always be measured by this essential criterion: the participation in and the access to the archive, its constitution, and its interpretation."[14]

Linked to authority and the establishment of power, archives also carry with them the threat of violence: a promise is also a threat.[15] Derrida, drawing from the etymology of the term *archive*, underscores that the archive is always both institutive and conservative: "It keeps, it puts in reserve, it saves, but in an unnatural fashion, that is to say in making the law (*nomos*) or in making people respect the law . . . it has the force of law."[16] The archive is, he argues, *toponomological*: it "coordinates two principles in one: the principle according to nature or history, *there*, where things *commence*—physical, historical, or ontological principle—but also the principle according to the law, *there* where men and gods *command*, *there* where authority, social order are exercised."[17]

These questions of violence and authority, of the transformation of a private to a public space, are key to understanding and assessing changes brought about by new media. New media have made certain archives more accessible by increasing the "domiciles" in which they—or copies of them—can be kept, spreading democracy by compromising privacy. At the same time, they have also made these files more volatile in both content and form, complicating the task of both preserving and reserving them. The flame wars and conflict online combined with increasing corporate and state surveillance—that is, our constant state of crisis—are not accidental to, or an unfortunate and temporary side effect of, new media; rather, they are symptomatic of changes in archival power, changes that underscore how new media can be the "end" of the archive, in both senses of that word. To return to new media as metaphor, the new media makes the archive metaphorical not only because, as Wolfgang Ernst has shown, it is not an archive, but also because it brings archives alive, shot through with change. Traces do not simply degenerate at a faster pace, they also transform themselves. This transformation challenges the process of consignation—of indexing and organizing—that grounds the archive; it also fundamentally changes how archived materials are retrieved, or "reanimated" and thus experienced. Perhaps, as Ernst has so eloquently argued, it is time to archive the word *archive*, assuming, that is, that we could separate its verb and noun forms.[18] Perhaps our very sense of the end of archives—their promise and their threat—is embedded in their structure as undead, as traces of the past we simultaneously integrate and forget as they endure and disappear.

3 Order from Order, or Life According to Software

We ought to regard the present state of the universe as the effect of its antecedent state and as the cause of the state that is to follow. An intelligence knowing all the forces acting in nature at a given instant, as well as the momentary positions of all things in the universe, would be able to comprehend in one single formula the motions of the largest bodies as well as the lightest atoms in the world, provided that its intellect were sufficiently powerful to subject all data to analysis; to it nothing would be uncertain, the future as well as the past would be present to its eyes. The perfection that the human mind has been able to give to astronomy affords but a feeble outline of such an intelligence.

—Pierre-Simon Laplace[1]

To repeat, software is axiomatic. As a first principle, it fastens in place a certain logic of cause and effect, a causal pleasure that erases execution and reduces programming to an act of writing. An axiomatic, as Gilles Deleuze and Félix Guattari contend, artificially limits decodings, it "blocks all lines [of flight], subordinates them to a punctual systems, and halts the geometric and algebraic writing systems that had begun to run off in all directions."[2] Software seeks to limit what can be decoded through an artificial boundary of programmability that renders hardware logical and orderly. This causal pleasure both stems from and bleeds elsewhere, making software— itself based on metaphor—a metaphor for the mind, for genes, for culture, for the economy, indeed for metaphor itself. Paul Edwards among others has outlined the ways in which the computer as a software/hardware machine conceptually grounded cognitive psychology, and it seems impossible to discuss biology without recourse to computer technology (to circuits, information exchange, and programs).[3]

This intertwining of computer technology and biology is hardly new, although it is constantly discovered as though it is. The history of computing and the history of biology are littered with moments of deliberate connection and astonished revelation, from computer storage as "memory" to regulatory genes as "switches," from genetic to evolutionary "programs." Through generative misreadings, biology and computer technology are constructed as complementary strands of a constantly unraveling and

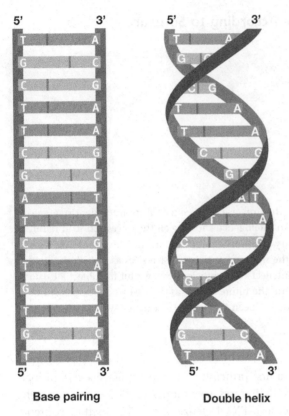

Base pairing   Double helix

Figure 3.1
Stylized double helix

raveling stylized double helix (see figure 3.1). That is, they are reduced to shiny corresponding balls or strips through the sleek simplicity of programmability, a force of attraction that binds the two together and enables them to reduplicate—to write each other—even as they work in opposite directions.

Richard Dawkins's "transformation" of evolutionary biology through the rubric of selfish genes exemplifies this logic: "we are," he writes, "survival machines—robot vehicles blindly programmed to preserve the selfish molecules known as genes."[4] Comparing neurons to transistors, he asserts, "brains may be regarded as analogous in function to computers."[5] This analogy makes clear the parallels not only between computers and humans, but also between them and a capitalist economy in which agents operate selfishly; it further naturalizes this comparison by making "nature" responsible for these actions. This analogy, which Dawkins claims still "fills [him] with astonishment," is also astonishing for its historical blindness, for Dawkins's revelation repeats John von Neumann's foundational analogy between neurons and computer components, as well as Nobel Prize winning geneticist François Jacob's 1970 claim that the involuntary aim of organisms is the reproduction of "an identical programme for the following generation. The aim is to reproduce."[6] In the first published account of modern stored program computing, the notorious 1945 "First Draft of a Report on the EDVAC," von Neumann asserts: "the three specific parts CA [central arithmetic], CC [central control] (together C) and M [memory] correspond to the associative neurons in the human nervous system. . . . The equivalents of the *sensory* or *afferent* and the *motor* or *efferent* neurons . . . are the *input* and *output* organs of the device."[7] Importantly, von Neumann's neurons are already cyberneticized: they are, as chapter 4 elaborates, the idealized neurons of Warren McCulloch and Walter Pitts, who used "neurons" to instantiate Turing's universal machine.

Each instance of connection, though, is repetition with a difference—one that creates a new linkage down the chain, but also poses the possibility of breakage or mutation. Dawkins's formulation, though, does not simply repeat Jacob's because it emphasizes the acts of individual "selfish" genes rather than the genetic program as a whole. It also does not simply repeat von Neumann's because it privileges a concept foreign to von Neumann in 1945: software. Although von Neumann details the "code" to be included in the memory in "First Draft," most of the report is devoted to outlining the function of logical hardware devices. Software had not emerged by then as a separate quantity; as outlined in chapter 1, the terms *program* and *code* in 1940s computing were mainly verbs. Again, writing in 1947 with Herman Goldstine, von Neumann insisted that coding was "not a static process of translation, but rather the technique of providing a dynamic background to control the automatic evolution of a meaning."[8] In von Neumann's analogy, code or program was not the pivot between the machinic and the biological.

As many have pointed out, the engineers who built the first digital computers did not foresee software; it was not foreseen, I wager in this chapter, because the drive for software—for an independent program that conflates legislation with execution— did not arise solely from within the field of computation. Rather, code as logos existed elsewhere and emanated from elsewhere—it was part of a larger epistemic field of biopolitical programmability. Part I of this book focuses on gendered and military relations; this chapter examines the intertwined and intertwining importance of Mendelian genetics and biopower more generally.[9]

To make this point, this chapter perhaps perversely reads Erwin Schrödinger's *What Is Life?*—widely though controversially considered to be the ur-text of modern genetics—as *What Is Software?* That is, *What Is Life?* epitomizes the drive for a code-based causality detailed in the first two chapters, a drive that software and not DNA would (*and only could*) instantiate. It also resuscitates the possibility of an all-knowing sovereign: a Laplaceian subject who could, by reading our genetic code-script, know the results of all causal connections. Untangling the ways that software and heredity intersect to create what François Jacob would call the "logic of life"—a logic that reduces action to word, life to a programmable code—this chapter argues that early to mid-twentieth-century genetics and eugenics prefigured the emergence of software. Both Mendelian genetics and software are based on the return to a Laplaceian universe in which order stems from order. Both are a return to a reductionist, mechanistic understanding of life, in which the human body becomes an archive. Part of the goal of this chapter then is to complicate the standard narrative within the history of science that biologists adapted the notion of a program from computer science—a narrative that rather remarkably treats software as though it has always existed. This chapter supplements the accounts of cybernetics' intrusion into the biological sciences by underscoring the difference between first-order cybernetic control and programmability while still acknowledging their commonality.

Software: Not Always Already There

Software has become such a powerful conceptual tool that it is hard to remember that it did not always exist. The privilege accorded to software as always already there is remarkable, especially in terms of theorizing biological phenomena. For instance, to explain the importance of experiments on bacterial cultures, Jacob has argued:

> Everything . . . leads one to regard the sequence contained in genetic material as a series of instructions specifying molecular structures, and hence the properties of the cell; to consider the plan of an organism as a message transmitted from generation to generation; to see the combinative system of the four chemical radicals as a system of numeration to the base four. In short, everything urges one to compare the logic of heredity to that of a computer. Rarely has a model suggested by a particular epoch proved to be more faithful.[10]

Explaining Schrödinger's 1943 formulation of a genetic code-script, Jacob similarly states, "heredity [in Schrödinger's text] functions like the memory of a computer."[11] The ahistoricity of this analysis is remarkable, for Schrödinger could not have had this comparison in mind. The first stored program computer, outlined in von Neumann's draft (itself drawing from the human nervous system), was published years after Schrödinger's address. In 1943, with the exception of Konrad Zuse's little-known digital computer in Germany, all functional nonhuman computers were analog; FORTRAN, the first widely used higher-level programming language, was completed in 1957.

This emphasis on software and computers is also evident in the critical and historical literature. Historian Lily Kay, for instance, has insightfully argued that molecular biology underwent a "gestalt switch to information thinking" in the 1950s.[12] This informational discourse, fueled by the metaphorical adoption of concepts from the communication technosciences (cybernetics, information theory, and computers), displaced the rhetoric of "biological specificity," which was dominated by mechanical

lock-and-key analogies. After the 1950s, genes transferred "information" and the cor
relation between nucleic acids and proteins catachrestically became a "code." Kay's
Who Wrote the Book of Life? A History of the Genetic Code details this rhetorical transformation and its fruitful misreadings of terms such as *information* and *code* in order
to explain how bodies became reduced to messages. In making this argument, she also
uses software as a theoretical tool. When describing "the conceptual and semiotic
impact of cybernetics," she argues, "such [closed feedback] mechanisms were not
confined to hardware; they also served as software for social technologies. Philoso-
phers and social theorists since the eighteenth century have visualized political, economic, and physiological stabilizations and correctives through models of closed
cycles (reminding us of the diachronic and synchronic nature of metaphors)."[13] In
this passage, Kay uses software as a theoretical construct to describe a historical reality.
Software becomes analogous to rhetoric itself.

This notion of software as rhetoric is further developed by theorist Richard Doyle
in his important analysis of the postvital body in *On Beyond Living: Rhetorical Transformations of the Life Sciences*. In it, he states, "the object of biology has somehow been
displaced, with the molecule overtaking or territorializing the organism and getting
plugged into the computer."[14] That is, DNA has become the sovereign source of life,
and the surface and depth of the organism imploded into a new density of coding.[15]
To make this argument, Doyle emphasizes the importance of language: "rather than
a mere description or heuristic for the life sciences, the rhetorics of code, instruction,
and program materialized beliefs into sciences and technologies."[16] Language, he
stresses, "serves as the repository of the unthought of science, its 'software.'"[17] He
coins the term *rhetorical software* to "foreground the relational and material interactions that make possible the emergence of scientific statements." Rhetorical software,
he explains, "while highlighting the textuality of scientific practices . . . avoids a
textual determinism: as any user of software knows, software is usable only within a
network of hardware and—this is frequently overlooked—'wetware.'"[18] Like many
metaphorical uses of software, Doyle's draws on software's ability to turn words into
actions; it does, however, also insist on a richer understanding of software, like rhetoric
itself, as dependent on context, on the audience and institutions. That Kay's and
Doyle's texts are two of the most brilliant and insightful on the intersections between
cybernetics and genetics is not irrelevant: in order to think genetics theoretically, it
seems necessary to assume software, and to assume a line of influence moving from
computers to biology.

This assumption that software is always already there is embedded in the very logic
of software itself, which grounds, as chapter 4 contends, the current conflation of
memory with storage. Software, in other words, is a device that makes possible the
"always already there": new media as somehow more lasting and more ubiquitous
than other media. What, however, happens when we treat software as a historical as

well as a theoretical device and "code" within the field of computing too as catachresis? What if we apply Doyle's insightful analysis back to itself and treat "software" (as Doyle treats language) as the "unthought" of his argument?

Importantly, cybernetics itself is based on a comparison between animals and machines: it is the study of control and communication in both. Norbert Wiener coined the term *cybernetics* in the 1940s, basing it, as mentioned previously, on the Greek term *kybernete* ("steersman" or "governor"). Wiener, assumed by many (including himself) to have founded the field, brought together work on electronic control systems—in particular negative feedback control—with studies of animal behavior.[19] Treating an animal's nervous system as analogous to electrical circuits, cybernetics modeled both as relaying messages (signals) that controlled action. Elaborating this analogy further in *The Human Use of Human Beings*, his popular follow-up to his 1948 work *Cybernetics, or, Control and Communication in the Animal and the Machine*, Wiener states:

> The physical functioning of the living individual and the operation of some of the newer communication machines are precisely parallel in their analogous attempts to control entropy through feedback. Both of them have sensory receptors as one stage in their cycle of operation; that is, in both of them there exists a special apparatus for collecting information from the outer world at low energy levels, and for making it available in the operation of the individual or of the machine. In both cases these external messages are not taken *neat*, but through the internal transforming powers of the apparatus, whether it be alive or dead. The information is then turned into a new form available for the further stages of performance. In both the animal and the machine this performance is made to be effective on the outer world. In both of them, their performed action on the outer world, and not merely their *intended* action, is reported back to the central regulatory apparatus.[20]

As this passage makes clear, cybernetics—although it insists that messages are "coded" (i.e., signals)—does not necessitate software: Wiener could easily be discussing analog computers, such as Bush's differential analyzer. The source of the animal or machine's actions is not some "programme," and messages are signals to be sampled and transformed, not things that drive the machine.[21] Although the military communications milieu was important to cybernetics' conflation of control with communication, this conflation occurs not necessarily or initially because a command given is a command executed, but rather because communications systems are control systems: control operates by comparing output with input.[22] In cybernetics as first conceived, there is no separation between software and hardware, an impossible distinction during the 1940s; Wiener's algorithms and instructions consist of relays.[23] What links the dead and the alive—perhaps making them both undead—is feedback.

Claude Shannon's (and Warren Weaver's) influential 1949 definition of *information* also emphasizes communication. Information gauges what can be communicated; it is "a measure of one's freedom of choice in selecting a message," not something that drives action.[24] As Shannon states, "the fundamental problem of communication is that of reproducing at one point either exactly or approximately a message selected at another point . . . the significant aspect is that the actual message is one *selected from a set* of possible messages. The system must be designed to operate for each possible selection, not just the one which will actually be chosen since this is unknown at the time of design."[25] Communications engineers use information to determine the necessary capacity of a channel, bus, or decoder. Thus the question unanswered by cybernetics' influence on the life sciences is: how did DNA (and indeed software) arise as the source? How was information transformed not only within the biological, but also within the computational, sciences? Taking into consideration the fact that initial computer punch cards, based on the Jacquard Loom, also worked via a physical lock-and-key mechanism, how did computer codes become language?

To unpack these questions, I begin with another: what happens if we take seriously Jacob's claim that Schrödinger's idea of heredity coincides with computer memory, years before such memory was developed? In other words, what if the text lauded as launching modern genetics—because it postulated the existence of a genetic code-script and because it inspired physicists such as Francis Crick and James Watson to move into biology from physics (even though the text was widely considered to be scientifically inaccurate or repetitive even at the time of its writing)—also inadvertently "launched" modern stored-program computers? What if the arrow of influence is not so simply one-sided (program moving from computers to biology), but rather part of a complex network of relations and productive, resonating misreadings? As I elaborate in more detail later, to reverse this arrow is not to imply a direct relationship between Schrödinger and von Neumann. Rather, it is to place both within a larger epistemic and governmental drive to make sense of the visible through an invisible program that links past to the present and that links individual to a population. As Jacob notes, the idea of reproduction as a combination of both parents, initially developed in the eighteenth century, necessitated the intervention of a "memory," a hidden, third-order structure that guided heredity.[26] With modern genetics, changes in this memory stemmed not from an invisible all-knowing "mysterious hand," but from the vagaries of population dynamics—what matters are not simply individuals, but rather the relations between them; it is these relations that elucidate invisible programs.[27] Both computer as memory machine and genetics as program are thus part of what Richard Panek has provocatively called "the invisible century," a move within the sciences in the twentieth century away from studying what is simply visible and experientially based—such as organization—toward speculation, toward theorizing "that there is more to the universe *than meets the eye.*"[28] Intriguingly, both Schrödinger and Jacob described the "governance" within cells in industrial or military terms: for Schrödinger, cells acted like an ideal military, in which each soldier had a full set of orders; for Jacob, bacteria acted like chemical factories.[29]

What Is Software?

In *What Is Life?*, Schrödinger, writing as an amateur biologist or "naïve physicist," outlines the challenge human genetics poses to then-current chemical and physical knowledge: given that statistical physics shows that Newtonian order only exists at large scales, how is it that the barely microscopic chromosomes guarantee the orderly succession of human traits? This question is, of course, one that only a modern physicist, familiar with Mendelian genetics, would ask, and this text—like Norbert Wiener's 1948 *Cybernetics, or, Control and Communication in the Animal and the Machine* —grapples with the ways in which life seems to defy the predictions of modern statistical physics. Life, Schrödinger writes,

> seems to be orderly and lawful behaviour of matter, not based exclusively on its tendency to go over from order to disorder, but based partly on existing order that is kept up. To the physicist—but only to him—I could hope to make my view clearer by saying: The living organism seems to be a macroscopic system which in part of its behaviour approaches to that purely mechanical (as contrasted with thermodynamical) conduct to which all systems tend, as the temperature approaches absolute zero and the molecular disorder is removed.[30]

More particularly, the regularity guaranteed by chromosomes defies fundamental assumptions, for at such a microscopic level, chromosomes should be destabilized by the impact of heat motion from surrounding molecules. They should act statistically, rather than mechanically. Given that they do not, Schrödinger argues, "the arrangements of the atoms in the most vital parts of an organism and the interplay of these arrangements differ in a fundamental way from all those arrangements of atoms which physicists and chemists have hitherto made the object of their experimental and theoretical research."[31]

Importantly, although he argues that the regularity of heredity is evidenced by certain constant features such as the Hapsburg lip, Schrödinger sees heredity as encompassing far more than the transmission of certain permanent traits. "We must not forget," Schrödinger writes,

> that what is passed on by the parent to the child is not just this or that peculiarity, a hooked nose, short fingers, a tendency to rheumatism, haemophilia, dichromasy, etc. Such features we may conveniently select for studying the laws of heredity. But actually it is the whole (four-dimensional) pattern of the 'phenotype,' the visible and manifest nature of the individual, which is reproduced without appreciable change for generations, permanent within centuries— though not within tens of thousands of years—and borne at each transmission by the material in a structure of the nuclei of the two cells which unite to form the fertilized egg cell.[32]

The visible regeneration of certain traits points to a larger invisible transmitted structure. To explain this "marvel" of regularity, Schrödinger argues that genes (which were then a hypothetical entity), and perhaps the chromosomes as a whole, must be an aperiodic crystal, for such a structure would be both stable and sufficiently complex to hold the entire "pattern" of an individual. Just as Morse code enables a complex message to be stored using just two signals—the dash and the dot—the gene could create a great many numbers by employing a simple number of signs. Since all crystals in the nonliving world are periodic, these structures would be peculiar to living organisms, but, importantly, they would still follow the laws of quantum mechanics: they would mutate into different isomers with the absorption of different quantae of energy.

Even if following quantum mechanics, living organisms—organisms that keep "'doing something' for a much longer period than we would expect of an inanimate piece of matter to 'keep going' under similar circumstances"[33]—Schrödinger postulates, must have some other source than energy. Indeed Schrödinger speculates that, in order for living organisms to create order from order, they "feed" on what he calls "negative entropy." That is, the living organism delays "the rapid decay into the inert state of 'equilibrium' [death]" by digesting "extremely well-ordered state of matter in more or less complicated organic compounds."[34] This idea, which resonates with Shannon's conception of information as negative entropy (a measure of randomness) and with Wiener's of information as positive entropy (a measure of order or structure), has led many to see it as foreshadowing the concept of information itself.[35]

The living organism—through its consumption of negative entropy and its embedded code-script—enables a return to the all-penetrating mind postulated by Pierre-Simon Laplace: "Every complete set of chromosomes contains the full code. . . . In calling the structure of the chromosome fibres a code-script we mean that the all-penetrating mind, once conceived by Laplace, to which every causal connection lay immediately open, could tell from their structure whether the egg would develop, under suitable conditions, into a black cock or into a speckled hen, into a fly or a maize plant, a rhododendron, a beetle, a mouse or a woman."[36] This return to Laplace is also a return to the possibility of a sovereign subject capable of "knowing all": if liberal governmentality, as stated previously, based itself on the repudiation of such a position, Mendelian genetics—through its linking of the visible to the invisible via the concept of code-script—signals a different relationship between individual and society. Not accidentally, this cellular system is also framed by Schrödinger as an ideal system of governance: in our bodies, every single cell has the instruction code for every other cell. He compares our cells to the "intelligent and reliable" soldiers engaged in General Montgomery's African campaign, all of whom were allegedly "meticulously informed of all his designs."[37] This, Schrödinger explains elsewhere, made the cell-state utopian: "A society which is organized on the principle that every official, every civil-servant, every person who has any duty at all within that organization is, at least in principle, given the same universal training and is so well informed about the plan of the whole, that every clerk could, in principle take over the duties of the prime-minister, every police-man that of a chief-surgeon, etc., etc."[38] Schröding-er's ideal is the mirror image of the utopia of Engelbart, Bush, and Hopper, in which the machine/code automates the position of the clerk, leaving the executives in place. In either case, however, one has an empowered individual agent capable of governing through enhanced knowledge.

Part of the "ideality" of Schrödinger's system stems from the fact that execution is inscribed within these instructions. For Schrödinger, the code-script was not only a plan but also execution. It was "the entire pattern of the individual's future development and of its functioning in the mature state."[39] Because of this, Schrödinger argues that the term *code* is inadequate: "the term code-script is, of course, too narrow. The chromosome structures are at the same time instrumental in bringing about the development they foreshadow. They are law-code and executive power—or, to use another simile, they are architect's plan and builder's craft—in one."[40] To refer back to chapter 1, code-script is code as logos; it is Lessig's "code as [automatically executing and regulating] law," a knowledge/action that distorts liberalism's blind "game."

This notion of code as both architect's plan and builder's craft in one clearly draws from, even as it mutates, the seventeenth-century notion of nature as the Book of Life. As Kay points out, few molecular biologists, even as they viewed life as a book, viewed God as its author.[41] Rather, as this notion of code as both execution and legislation makes clear, the writing becomes the writer; it becomes a powerful source that also teases us with the possibility of sovereign power, with the possibility of editing, rewriting, and creating new genetic codes.[42] Describing the power of natural selection, Jacob contends, "without any thought to dictate it, without any imagination to renew it, the genetic programme is transformed as it is carried out."[43] This sovereign power, sans Sovereign, makes the world infinitely malleable: what natural selection unconsciously accomplishes, humans can do deliberately. Tellingly Jacob, describing his embrace of atheism, writes, "If God did not exist, it was necessary to do without him. An empty heaven left an earth to fill, and it was up to me to fill it. A world to construct, and it was up to me to construct it."[44] Jacob's sentiment resonates with comments made earlier regarding the power and pleasure of programming—programming as a form of pleasurable megalomania.

Assessments of the importance or accuracy of Schrödinger's text vary widely—from biologists Watson and Crick, who have listed it as a direct inspiration; to chemist Linus Pauling, who viewed it as retarding the development of molecular biology by leading biologists to concentrate on life as entropy;[45] to historian of science Lily Kay, who has argued that Schrödinger's concept of "pattern" does not correspond to modern molecular biology, since it is aligned with the older, protein view of life.[46] According to historian Donald Fleming, *What Is Life?* is "a book that was in league with the future but scientifically antiquated before it was written."[47] The only thing all the commentators agree on—and are perhaps puzzled by—is its enduring popularity. As phage biologist

Gunther Stent puts it, "just why this book should have made such an impact was never quite clear. After all, in it Schrödinger presented ideas that were even then neither particularly novel nor original." By bringing biology to the attention of physicists, however, Stent contends, *What Is Life?* became "'the Uncle Tom's Cabin' of the revolution in biology that, when the dust had cleared, left molecular biology as its legacy."[48] Relatedly, Fleming argues that *What Is Life?* was emblematic of the impact of "émigré physicists," whose move into biology inspired bold reductionist projects about "the secret of life."[49] In contrast to the biologists who, steeped in reverence for biological specificity, issued cautious statements, physicists proffered grand simplifications and statements. According to Evelyn Fox Keller, this push toward all-encompassing rhetoric did not only stem from methodology: physics and physicists, such as Schrödinger, supplied biology with social authority and social authorization, enabling biology to borrow physics' agenda, language, and attitude, and even its names.[50] In response to these conflicting interpretations, Leah Ceccarelli theorizes that *What Is Life?* negotiates these different interests and beliefs through a strategic ambiguity that makes all these readings possible.[51] Ceccarelli's reading is convincing, especially if one considers Schrödinger's description of "code-script." Even if Schrödinger's concept of protein is aligned with the older protein view of life, the term *Morse code* implies the transfer of information. According to Ceccarelli's rhetorical reading, Schrödinger's text reveals the value that untrue, unoriginal—"vapory"—science can have.[52]

The value of *What Is Life?*, nevertheless, varies significantly with the "future" from which *What Is Life?* is evaluated. Again, from the perspective of chemists or of those assessing the impact of Schrödinger's text from molecular biology's later turn away from physics toward chemistry, the value of *What Is Life?* is negative, or at the best inspirational (Schrödinger himself argued that biochemists, not physicists, were going to be responsible for the next advances in the study of heredity). For those working from the perspective of late twentieth-century genomics, focused on cracking the code of life, Schrödinger's text is central. Doyle, for instance, contends that *What Is Life?* made possible a "post-vital body," "a body in which the distinct, modern categories of surface and depth, being and living, implode into the new density of coding, what Jacob calls the 'algorithms of the living world.'"[53] That is:

> No longer a reflection or even a production of genotype, "pattern" is now literally inside genotype. By "troping" the trope of pattern, Schrödinger literally and grotesquely turns "pattern" and the "organism" inside out. With this move—the metonymic substitution of "code" for "organism"—the entire future birth, life and death of the organism is "contained" or engulfed by the chromosomes. . . . Schrödinger mistakes or displaces the pattern of the organism by its "code-script," injecting the life of the organism into its description . . . Schrödinger places all the power within the code and none within the development of the organism.[54]

This rhetorical move of "troping the trope," Doyle claims, also made it thinkable, practicable, for Watson and Crick to claim that "decoding" the structure of DNA equals decoding life. With the "injection of 'law code and executive power' into DNA, *code* becomes as much a verb as noun, the double helix becomes as much body as its description."[55] Although Doyle is careful to point out that DNA as life is only possible "after the articulation of the structure of DNA and the tropics of 'code' get played out," his description does overlook the question of those other laws of nature that Schrödinger posits as necessary to *how* genes and chromosomes operate.[56] To Schrödinger, the main characteristic of life again was not the code, but the fact that the living organism "kept moving."

Stepping aside from debates over the value of *What Is Life?*, I want to use Doyle's argument to outline another debate: the relevance of this text, and modern conceptions of heredity in general, to the emergence of software, for Schrödinger's positing of a code that is both law and execution arguably foreshadows code as computer. Alexander Galloway's notion that hardware does not do anything, addressed in chapter 1, itself depends on the tropics of code. Code-script is source code as logos. It axiomatically, temporarily, limits the entropic nightmare of decay that looms as an absolute limit to capitalist progress through a privileging of causality that stems in part from an acceptance of Mendelian laws of heredity—the fascination of the stability of heredity reduced to the constancy of code. In other words, to understand the impact of Schrödinger's text on biology, we need to look at its coincidence with computer technology. We need to look at the ways in which genetics has put in place, while also drawing from, dreams of programmability, dreams that computer technology and not biology would, and could, only come to instantiate.

Again, I am not arguing for a direct line, or secret meeting, between Schrödinger, John von Neumann, Alan Turing, and Grace Murray Hopper, for this argument seeks to break free from a logic that focuses solely on direct citations and that adjudicates the scientific validity of such borrowings. The connection I'm making is more general, almost archeological (in the Foucauldian sense). According to Michel Foucault, the archive is "first the law of what can be said, the system that governs the appearance of statements as unique events." The archive does not simply control the emergence of statements, but also "determines that all these things said do not accumulate endlessly in an amorphous mass, nor are they inscribed in an unbroken linearity, nor do they disappear at the mercy of chance external accidents; but they are grouped together in distinct figures, composed together in accordance with multiple relations, maintained or blurred in accordance with specific regularities."[57] Foucault emphasizes that these laws that "determine[] at the outset *the system of its functioning*" neither emanate from outside, nor from the object nor from an exterior idea (from what is not said), but rather are immanent in discourse itself.[58] The archive is what Foucault earlier calls the "positive unconscious of language" in *The Order of Things: An Archaeology of the Human Sciences*;[59] that is, the well-defined regularity of empirical knowledge, which is part of scientific discourses, even as it alludes to the consciousness of the scientist.

As a system of regularities or the law that drives diverse statements, the archive is not, however, aligned with continuity (and, thus, with traditional history)—it does not place us in a realm of preserved discourse—but rather with discontinuities. The archive, Foucault emphasizes, deprives us of the continuities necessary to establish a stable temporal identity:

> It breaks the thread of transcendental teleologies; and where anthropological thought once questioned man's being or subjectivity, it now bursts open the other, and the outside. In this sense, the diagnosis does not establish the fact of identity by the play of distinctions. It establishes that we are difference, that our reason is the difference of discourses, our history the difference of times, our selves the difference of masks. That difference, far from being the forgotten and recovered origin, is this dispersion that we are and make.[60]

Thus, regularity and subjectivity, for Foucault, are found in dispersion rather than in continuity or in stability. Discontinuity is key and this emphasis on discontinuity encapsulates archaeology's difference from traditional history as the history of (and as enabling) the continuity of human consciousness. Indeed Foucault describes *Archaeology of Knowledge* as an enterprise to "measure the mutations that operate in general in the field of history."[61] Gilles Deleuze, summarizing Foucault's archeological project, claims that each age has its own particular distribution of the visible and the articulable.[62] The archeological project attempts to map what is visible and what is articulable, and to understand how visibilities and language operate.

Crucially, the archeological project itself is not separate from the historical formation that makes possible *What Is Life?* as *What Is Software?* (or, in more Foucauldian terms, that makes them the same statement); Foucault's archive—his search for fundamental codes and laws—is deeply intertwined with these projects. That is, the desire to map what is visible and what is articulable is key to understanding the impact of code and programmability—to the linking of the two "programmed visions." Programmability is thus not only crucial to understanding the operation of language, but also to how language comes more and more to stand in for—becomes the essence or generator of—what is visible.

Foucault's reliance on notions of law and discontinuity, and the terms he uses to describe them, resonate with mathematics and cybernetics. The archive, he argues, is a "complex volume," and order is one of the "fundamental codes of a culture."[63] The archeological project takes on this question: what "if errors (and truths), the practice of old beliefs, including not only genuine discoveries, but also the most naïve notions, obeyed, at a given moment, the laws of a certain code of knowledge?"[64] Later in his writings, he describes power in terms of a "network" and archeology itself as a grid, which places power on one axis and knowledge on the other.[65] Intriguingly, he describes statements, the alleged fundamental elements of the archive, not as atomic units, but rather as an enunciative function that involved various units . . . and instead of giving a "meaning" to these units . . . relates them to a field of objects; instead of providing them with a subject, it opens up for them a number of possible subjective positions; instead of fixing their limits, it places them in a domain of coordination and coexistence; instead of determining their identity, it places them in a space in which they are used and repeated. In short, what has been discovered is not the atomic statement—with its apparent meaning, its origin, its limits, and its individuality—but the operational field of the enunciative function and the conditions according to which it reveals various units.[66]

This relation of what can be seen and what is not hidden yet driving—and which is not terminal—coincides with our perception of the relationship between a program and its interface. This is not to say that Foucault views statements as "source code"; this is the opposite of his approach. This is to say that this notion of an operational field of enunciative function resonates with von Neumann's notion of code as a dynamic "context," as something that does not pin down a meaning, but rather guides—makes possible—certain calculations.

In other words, archeology, even as it admits to and emphasizes discontinuity and dispersion, also seeks to make causal relations between what is seeable and what is seen. This rhetoric and these coincidences are not accidental, but crucial to connecting the lines between Foucault's project and the twin projects of genetics and computer technology. This chapter therefore does not simply "discover" the archeological link between computer technology and genetics, but rather contends that computer technology, genetics, and archeology are part of the same archive. They similarly rely on, they create, discontinuous rather than continuous knowledge and disperse knowledge along parallel trajectories. As I elaborate later, we can arguably now recognize these similarities, make sense of or outline this particular archive, because knowledge is moving elsewhere: the archive is only visible as it recedes from us. As Foucault puts it, "it is the border of time that surrounds our presence . . . the description of the archive deploys its possibilities (and the mastery of its possibilities) on the basis of the very discourses that have just ceased to be ours."[67] Foucault importantly supplemented archeology with genealogy and strategy. Quantum computing is closer to analog rather than discrete computing; there is no one-to-one relation between genes and enzymes; in fact the same DNA sequence can code more than one protein. Due to increased research on retroviruses such as AIDS, RNA, rather than DNA, is more and more considered the source of life.[68] That is, as chapter 4 argues, rather than producing possible actions and statements, the archive is a constantly rewritten storage system, driven by the ephemeral.

The Returns of Laplace

Tellingly, this return to Laplace takes place in an equally classic, if far less controversial, text in the field that would become computer science: Alan Turing's 1950 "Computing

Machinery and Intelligence," in which he describes discrete state machines as universal mimics.[69] In discrete state machines, it is always possible to predict all future states, given the initial state of the machine and the input signals. This is reminiscent, he writes, of "Laplace's view that from the complete state of the universe at one moment of time, as described by the positions and velocities of all particles, it should be possible to predict all future states." Discrete state machines, however, enable a return to Laplace through a simplification of mechanics. In "the system of the 'universe as a whole' . . . quite small errors in the initial conditions can have an overwhelming effect at a later time. The displacement of a single electron by a billionth of a centimeter at one moment might make the difference between a man being killed by an avalanche a year later, or escaping." By contrast, in the mechanical systems Turing explicates, this phenomenon does not occur. Even when we consider the actual physical machines instead of the idealized machines, reasonably accurate knowledge of the state at one moment yields reasonably accurate knowledge any number of steps later.[70] In discrete state "mechanical" machines, that is, order follows from order. Significantly, in this passage, Turing is referring to hardware rather than to software—or, to be more precise, he is not dealing with the separation of software from hardware, which would take place much later, under commercial pressure. Discrete state machines are predictable because hardware is used in particular ways: gates are carefully timed so that delays do not produce significant false positives or negatives, signals are rectified so that they can be read correctly, hardware is carefully built to cut down on voltage spikes and crosstalk, and so on. Software is axiomatic, but only because our discrete hardware is constructed to be so. In contrast to analog computing, which sets the conditions and parameters for the analogy to run, discrete computing uses numerical methods, which demand step-by-step control and accuracy. Discrete hardware, in other words, is constructed to follow (and store) instructions. As I argue in the next chapter, digital logic, which makes software possible, is axiomatic.

Although hardware grounds universality (or something close to universality, that is, digitality), software, as an entity independent of hardware, is crucial to machines as "self-reproducing," that is, key to Jacob's reading of Schrödinger's notion of heredity as memory. In fact, von Neumann, as chapter 4 elaborates, turns from his usual cybernetic parallel between computing machines and the human nervous system toward genetics and programs when addressing the question of self-reproducing automata. As Kay notes, von Neumann's interest in biology, "in general, and genetics, in particular, became closely linked to his mission of developing self-reproducing machines."[71] In his outline of a general descriptive theory of these types of automata in his 1951 "General and Logical Theory of Automata," von Neumann postulates that such automata contain within themselves an instruction set I_D, which holds a description of itself. He argues: "It is quite clear that the instruction I_D is roughly effecting the functions of a gene. . . . It is, of course, equally clear at which point the analogy ceases to be valid. The natural gene probably does not contain a complete description of the object whose construction its presence stimulates. It probably contains only general pointers, general cues."[72] This move toward what will become code as logos is thus linked to genetic inheritance. It is intriguing that, to make this move, von Neumann turns to a biological concept as fuzzy as memory for his analogy (like software as metaphor, memory as metaphor explains an unknown through another unknown). Thus to supplement Kay's argument, it is not that von Neumann moves toward biological systems when he begins investigating self-reproducing automata—his description of the computer already compared biological and machine systems at the level of nerves and electricity—but that the move to self-reproduction and thus programmability marks a significant change in the ways in which both biological and machine systems are compared, namely at the level of what would become software. The question of self-reproduction, as posed by von Neumann, is largely a question of transmission. That the instructions can effect development is already assumed.

But why is this move to genetics key to code as logos? Why this intertwining of programmability—the production of constantly transmitted visible characteristics—with genetics and to what extent does it rely on a certain discrete logic? What is the relationship between the transmissible, the programmable, and the discretely invisible?

Invisible Transmissions, Visible Results

As noted earlier, Schrödinger himself assumes that the transmission of certain features, such as the Hapsburg lip, is not only evidence of the stable transmission of that feature, but also of human heredity in general (according to his biographer, when informing his illegitimate daughter of her paternity, he pointed to their similar feet). This notion of a discrete trait as evidence for larger processes of human heredity—as a marker of history—is linked to the "rediscovery" of Mendel, itself an event, theorized by Vannevar Bush in his essay "As We May Think," as revealing the need for better means of information archival and retrieval. As the rest of this section argues, Mendelian genetics, with its emphasis on discrete units within a population and on the relationship between genotype and phenotype, provided a compelling model for the continuing relevance of mechanistic and reductionist understandings of nature. This mechanism and reductionism not only made possible the description of heredity, but also the possibility of effecting heredity, of a genetic program. That is, in order for Mendelian genetics to operate as a science, it had to be able to make predictions. These predictions in turn were predicated on active interference and experimentation in model species such as drosophila, if not humans. As geneticist Hermann J. Muller in an unappreciated Marxist analogy stated at the Second Annual International Conference on Eugenics: "Beneath the imposing structure called 'Heredity' there has been a dingy basement called 'Mutation.'"[73] Mutation meant, especially for Muller, actively mutating the animal's genotype. The methodology of breeding—especially breeding mutant lines—was thus fundamental to the establishment of genetics as a science.

Programmability—the past as determining the future—in other words, is not just the application of genetics, but also its proof and methodology. Mendelian genetics was proven as a true predictive science through successful breeding experiments, which visibly revealed invisible essences and mechanistic, causal laws. Eugenics was not simply an application of genetics, but rather proof of genetics' predictive power; early geneticists such as Muller were also eugenicists, and many early genetics texts such as Muller's address and R. C. Punnett's *Mendelism* explicitly link them together.[74] When eugenics fell into scientific disrepute in part because human genetics was so difficult to program, the desire for programmability became encapsulated within software and, as Lily Kay has outlined, within molecular biology. Mendelian genetics, by postulating a relationship between phenotype and genotype, also put into play a relationship between what is invisible and visible; the ambiguous distinction between the two would be generative for many years to come.

The historical events and controversy surrounding the "rediscovery" of Mendel have been well documented. Mendel's experiments were "rediscovered" independently by three different scientists at the end of the century, and each scientist, in presenting his findings, claimed Mendel as a lost predecessor. These claims of the mythic, lone true experimentalist ignored by the scientific establishment, however, are not limited to Hugo de Vries, Carl Correns, and Erich Tschermak von Seysenegg. As Jan Sapp has revealed, Gregor Mendel seems to have (at least) nine lives. They are:

1. Mendel was a non-Darwinian. Although he was an evolutionist, he did not entirely agree with Darwin's views and set out to disprove them. (Bateson 1909)
2. Mendel was a good Darwinian. His experimental protocols and reported results can be explained on the assumption that he had no objections to Darwinian selection theory. (R. A. Fisher 1936)
3. Mendel was not directly concerned with evolution at all. He placed it on the back burner while he investigated the laws of inheritance. (Gasking 1959)
4. Mendel rejected evolutionary theory. (Callender 1988)
5. Mendel laid out the laws of inheritance, which justifiably carry his name. (Standard view: see, e.g., Zirkle 1951, Mayr 1982)
6. Mendel was no Mendelian. He was not trying to discover the laws of inheritance, and several Mendelian principles are lacking in his papers. (Callender 1988; Brannigan 1979, 1981; Olby 1979)
7. Some of Mendel's data was falsified. (R. A. Fisher 1936)
8. None of Mendel's data was falsified. (See, e.g., Beadle 1966, Dunn 1965, Olby 1966, Wright 1966, Thoday 1966, Mayr 1982, Pilgrim 1984, Edwards 1987, Van Valen 1987)
9. Mendel's reported experiments set out in his paper of 1866 are wholly fictitious. (Bateson 1909)[75]

To explain these conflicting interpretations, Sapp claims that Mendel's place in history is not determined by his writings published in the 1860s, but rather through posthumous stories about Mendel and his experiments.[76] At stake in these interpretations, Sapp argues, "is a definition of the concepts and/or movements that can be legitimately associated with the genetics tradition," that is, Mendel as source.[77]

Crucially—although this seems a trivial point—all these interpretations agree that Mendel treated continuous traits discontinuously. Whether Mendel was a Mendelian and thus considered these traits to prove the laws of heredity or whether he worked in the tradition of plant hybridizers and thus focused on breeding "true" new species, Mendel isolated traits—such as wrinkled (or not), round (or not)—and treated these traits independently (a forerunner of the law of segregation). Following these discrete characteristics rather than the organism as a whole, or more clearly continuous and blended traits, allowed Mendel to focus on questions of stability rather than dynamic change. This focus, Daniel J. Kevles has revealed, contributed to Mendel's neglect, since it ran counter to the then-current thinking in evolutionary biology, which centered on modes of adaptation and change.[78] Mendel also differed from his scientific contemporaries in his focus on inter- rather than intraspecies (i.e., "racial") hybrids. Raphael Falk and Sahotra Sarkar argue that Mendel was responding to his Moravian countrymen farmers' concern with the return in their new breeds of what would become known as recessive traits. He thus sought to document the phenomenon of dominance in hybrids and to formulate the laws of the reappearance and disappearance of traits (regardless of whether or not he distinguished between phenotype and genotype).[79] Most pointedly, Punnett—an early adapter and advocate of Mendelian genetics—developed a theory of absence and presence, which made genetic traits fundamentally binary: it was not that dominant and recessive factors produced different characteristics, but rather that dominance marked the presence of a factor and the recessive its absence.[80]

This focus on discrete traits separated Mendelians, such as Punnett and William Bateson, from the more established biometricians, such as Francis Galton, Karl Pearson, and Walter Raphael Weldon, founders of British eugenics. The biometricians viewed evolution as working through small continuous changes. The main tool of Galton was the normal curve. Busily plotting graphs of various characteristics (such as height), Galton believed that the mean represented the center of a trait and that natural selection worked by moving this center elsewhere, although there was an overwhelming tendency to regress to some "racial center." Pearson, responding to the empirical fact that humans did not seem to regress in this manner and that small, seemingly statistically insignificant fluctuations could effect change, argued that the focus of regression was the immediately prior generation. Selective breeding could thus easily change this center.

In many ways, the biometrician's version of eugenics was more optimistic—or at least more inclusive—than that of the Mendelians, since there were no pesky recessive genes, which Mendelian eugenicists believed made a "melting pot" impossible. Although many biometricians supported the sterilization of extreme statistical outliers (Galton indeed coined the term *eugenics* as "the science which deals with all influences that improve the inborn qualities of a race; also with those that develop them to the utmost advantage") and although they separated transmission from development, their belief in the normal curve meant accepting those on either side of the norm as part of the same curve and thus their offspring as possibly "improvable," rather than as carriers of recessive traits that could forever damage their spawn.[81] The focus was not on the creation of pure lines. This curve, of course, was racialized—different races were represented by different curves.[82] This "optimism" regarding the improvability of the human race by the biometricians also did not translate into progressive politics. Both biometrician and Mendelian eugenicists were politically diverse: Galton was a conservative while Pearson a socialist; similarly, Charles Davenport a conservative while Muller a radical.

The biometricians, unlike the Mendelians, focused on what was visible. Even though the norm was an abstract principle, enumeration, as Kevles contends, did not mean penetrating beyond the phenomenological surface.[83] Heredity was a quantitative, correlative relationship between generations, not a causal one. As Galton explains, whereas "formerly the quantitative scientist could think only in terms of causation, now he can think also in terms of correlation."[84] For this reason, the biometricians were heavily involved in mass calculation. Mendelians too relied on statistical analysis, but for the early Mendelian geneticist, statistics were used to determine whether or not a trait was Mendelian. The seductiveness of early forms of Mendelianism lay in their mechanistic, causal laws.[85]

Classical Mendelian genetics separated genotype from phenotype—that is, what was transmissible from what was visible—as well as transmission from development (like the biometricians). This distinction between genotype and phenotype was first postulated by Wilhelm Johannsen, who experimented with inbred beans. "Natural" selection, Johannsen showed, did not effect these beans' length and breadth: the progeny of the beans consistently followed the same curve as that of their parents, proving the constancy of genetic materials. Johannsen's formulation of a genotype relied on treating a continuous characteristic—height—discontinuously, as a Mendelian feature that was passed on or not.[86] Johannsen's genotype, though, was not an actual entity. According to Nils Rolls-Hansen, his genotype was an ideal, inaccessible form, whose existence was posited through inference.[87] Although this concept seems very close to the biometricians' concept of a norm, Johannsen's distinction between genotype and phenotype suggested a new level of analysis and intervention for the Mendelians that followed him.[88] That is, rather than quantifying the visible surface, the

genotype implied that one had to penetrate the organism to understand the relationship between what could or could not be seen. Indeed, most Mendelians believed that
the trait itself was not transmitted, but rather the potential for the trait: something
that could make this trait visible, something invisibly visible or visibly invisible.
Johannsen argued that the genotype, rather than the phenotype, was transferred
between parent and offspring: according to his argument, both shared the exact same
genotype, making the genotype, like computer memory, a strangely ahistorical entity
nonetheless key for any historical relationship.[89]

Eugenics as Nurture

The understanding of genes as ahistorical also depends on the separation of transmission from development. It is a reduction of heredity to transmission, a clear separation
of nature from nurture, of germline from somatic cells. As George Stocking Jr. has
explained, the nineteenth century generally did not distinguish definitively between
race and nation:

> In 1896, the processes and the problems of heredity were little understood, and "blood" was
> for many a solvent in which all problems were dissolved and all processes commingled.
> "Blood"—and by extension "race"—included numerous elements that we would today call
> cultural; there was not a clear line between cultural and physical elements or between social
> and biological heredity. The characteristic qualities of civilizations were carried from one
> generation to another both in and with the blood of their citizens.[90]

Instead of race and nation standing for two different entities in the nineteenth
century, Stocking contends that they were separated by degree, with race implying a
greater degree of kinship.[91] Nineteenth-century views on race, in other words, included
what are now considered effects of culture; they were also far more Lamarckian, since
acquired characteristics were transmitted from generation to generation. Darwin, for
instance, condemned slavery as a major cause of the physical degradation of native
Africans. Darwin believed in "pangenesis," that is, a hereditary mechanism in which
every cell in the body played a part in forming the reproductive cells by shedding
"gemmules." According to William Provine, Darwin's attitude was heavily influenced
by animal breeders, who believed that virtually all physical and behavioral features
were partially hereditary.[92] As Stocking notes, this more Lamarckian view of race was
hardly less racist, and the subsequent separation of nature and nurture not only fostered more overt racism, but also more antiracist positions, such as that of Franz Boas,
who emphasized the importance of nurture over nature.[93]

The nature–nurture divide, however, also originated in the nineteenth century, with
Galton's refutation of what Ernst Mayr—in language clearly resonating with computer
technology—has called Darwin's theory of "soft inheritance."[94] Through an experiment with blood transfusions and subsequent inbreeding in rabbits, Galton showed

that the color of the offspring of the transfused rabbits never deviated from the parental color. Rather than all cells contributing gemmules, Galton theorized a "hard inheritance," in which the reproductive "stirp" (germ plasm transmitted from parents to child) was segregated from the rest of the body. August Weismann influentially drew from Galton to theorize the "continuity of the germ plasm," in which the "germ track" is separated from the "soma track" from the very beginning.[95] Eugenics—both biometrician and Mendelian—thus initiated the separation of nature (the hard) from nurture (the soft) by positing inheritance as outside experience, although again, what was considered hard versus soft is far different than it is today.

The standard position on the nurture–nature divide is, of course, that it coincides with a divide between what is and what is not under human control. Diane Paul offers the following concise summary of this standard position: "In the twentieth century, to hold that differences among human groups are biologically-based is necessarily to imply that those differences are largely outside of human control. . . . In this context, the epithet 'racist' has come to be applied almost exclusively to those views which ascribe non-trivial differences among human populations to biological, hence more or less permanent, differences."[96] This popular opinion, however, misses the point. The distinction between nature and nurture made possible by eugenics and by advances in human physiology, which showed that the reproductive cells segregated and formed at an early stage in human development, did not place nature outside of human control. The battle between nature and nurture was a battle over what type of control to use: eugenics or social welfare. In other words, the insightful observations by Evelyn Fox Keller and Eve Kosofsky Sedgwick that now (with the possibility of genetic engineering) nature is viewed as more flexible than nurture/culture needs to be pushed further, for the very positing of the nurture–nature divide established nature as an object of control and manipulation.[97]

Genetics separated cultural and biological transmission, but in doing so also made biological transmission a question of transmissible, cultural knowledge—a question of and for the archive. Although both the biometricians and the Mendelians separated the germline from somatic cells, the Mendelians' emphasis on invisible causal, mechanistic laws, rather than on visible statistical relationships, made their theories more pointed for theorizing the human body as archive. Punnett, for instance, insisted that the effects of education and hygiene were salutary but had to be renewed with each generation because they were not biologically inherited. He compared the human hoard of knowledge to a bee's store of honey: "each generation in using it sifts, adds, and rejects, and passes it on to the next a little better and a little fuller." This knowledge, however, is not an inheritance because

> the handing on of such knowledge has nothing more to do with heredity in the biological sense than has the handing on from parent to offspring of a picture, or a title, or a pair of boots. All these things are but the transfer from zygote to zygote of something extrinsic to the species. . . .

Better hygiene and better education, then, are good for the zygote, because they help him to make the fullest use of his inherent qualities. But the qualities themselves remain unchanged in so far as the gamete is concerned, since the gamete pays no heed to the intellectual development of the zygote in whom he happens to dwell.[98]

Although this would seem to construct biological transmission (nature) as trumping human control, it nevertheless places human evolution—as a matter of reproductive choice—within human control. Punnett goes on to state, "by regulating their marriages, by encouraging the desirable to come together, and by keeping the undesirable apart we could go far towards ridding the world of squalor and the misery that come through disease and weakness and vice."[99] Key to doing this, however, is more knowledge: "Before we can be prepared to act, except, perhaps, in the simplest of cases, we must learn far more about them. At present we are woefully ignorant of much, though we do know that full knowledge is largely a matter of time and means."[100] Similarly, Charles Davenport—the father of the U.S. eugenics movement and founding director of the Station for Experimental Evolution at Cold Spring Harbor, Long Island (now one of the most respected centers for the study of genetics)—ends his influential eugenics textbook *Heredity in Relation to Eugenics* by arguing for the necessity of a state eugenic survey. Eugenics is based on a fundamental belief in the knowability of the human body, an ability to "read" its genes and to program humanity accordingly.

The move from "they" to "we" in Punnett's statement is telling: eugenics is a collective program of controlling others and thus improving ourselves. With this "we," the knowledge of scientists and their capacities to intervene are conflated with those of society as a whole. Like cybernetics, eugenics is a means of "governing," or navigating nature. Similarly, Julian Huxley believed that the goal of eugenics was to control the evolution of the species and to guide it in a desirable direction.[101] Curt Stern, writing in the twentieth century in a textbook on human genetics claimed: "Natural selection will be superseded by socially decreed selection. In the course of time . . . the control by man of his own biological evolution will become imperative, since the power which knowledge of human genetics will place in man's hands cannot but lead to action. Such evolutionary controls will be world wide in scope, since, by its nature, the evolution of man transcends the concept of unrestricted national sovereignty."[102] Thus, it is not simply, as Garland Allen argues (in "The Social and Economic Origins of Genetic Determinism: A Case Study of the American Eugenics Movement 1900–1940 and Its Lessons for Today") that hereditarian thinking, stemming from economic and social conditions, distracts us from the social solutions before us by promising technological fixes, but also that eugenics and hereditarian-based political arguments more generally are themselves social solutions that demand further development of the human archive of knowledge.[103]

Eugenics, in other words, is a key component of what Michel Foucault in his *History of Sexuality*, volume 1, called biopower. According to Foucault, biopower is power focused on administering life or disallowing it to the point of death; it is "power bent on generating forces, making them grow, and ordering them, rather than one dedicated to impeding them, making them submit or destroying them."[104] Eugenics, according to Galton, was a science focused on improving stock; Galton "derived" eugenics from *eugenes*, the Greek term meaning "well born." Eugenics is biopower, situated and exercised, as Foucault insisted, "at the level of life, the species, the race, and the large-scale phenomena of population."[105] It is based on "the fundamental fact that human beings are a species,"[106] and it substitutes "population" for the rights and responsibilities of the subject or the sovereign as the "vis-à-vis of government, of the art of government."[107] Indeed, Davenport in *Heredity in Relation to Eugenics* debates whether or not an individual who commits premeditated murder is "responsible" for the crime. Putting himself in that situation, he argues, "I am not responsible in the sense of 'deserving' pain because of the inadequacy of the determiners in my protoplasm. . . . I am not responsible for my early culture nor for reactions determined by it; but that culture is partly determined by my makeup, as when I find pleasure in the society of bad companions, and partly is imposed by formal 'good influences' that society has organized."[108] This, however, is no excuse to add another burden to society, and so organized society must prevent the "automatic" effects of bad breeding through eugenics, "preventing the mating that brings together the antisocial traits of the criminal."[109] In Davenport's formulation, humans are both evaluated in terms of their economic costs to society and viewed as a form of capital: breeding must control both "innate" and acquired elements (innate qualities can be controlled through human, not technical, reproduction).[110] Sexuality knits the individual to the population and privileges the population—and its betterment—over the individual, bizarrely absolving him of both rights and responsibilities. Also, although moving responsibility from the individual to society in general, eugenics offers the individual the means to "map" an otherwise invisible system so that she can make the right marriage decisions.

Eugenics, like biopower more generally, focuses on sexuality—the early eugenics (and some early genetics) textbooks read like early twentieth-century marriage guides, educating its middle-class readers on how to marry wisely. The most powerful and long-lasting effect of the early eugenicist movement in the United States is the birth control pill. Margaret Sanger, an important birth control advocate, argued in 1920: "Birth control itself, often denounced as a violation of natural law, is nothing more or less than the facilitation of the process of weeding out the unfit, of preventing the birth of defectives or of those who will become defectives."[111] According to Foucault, biopower brings together a focus on the individual body (mechanisms of discipline) with a focus on the general population "through concrete arrangements," or technologies of power, of which the deployment of sexuality is the most important: "at the juncture of the 'body' and the 'population,' sex became a crucial target of a power organized around the management of life rather than the menace of death."[112]

Eugenics is clearly a means by which both individual and population were managed; it is also a means by which biopower could focus on questions of death in order to foster life. As a large-scale project, it is a manifestation of biopolitical governmentality, wherein governmentality is "a state of government that is no longer essentially defined by its territoriality, by the surface occupied, but by a mass: the mass of the population, with its volume, its density, and, for sure, the territory it covers, but which is, in a way, only one of its components."[113] In such a state, one that "takes life as both its object and its objective," eugenics and state racism become ways of giving the state the power of death. Racism, Foucault argues, "is primarily a way of introducing a break into the domain of life that is under power's control: the break between what must live and what must die." It introduces a difference—a discontinuity—that perpetuates a lethal difference. In biopolitical governmentality, the death of the other is also linked to the improvement of one's "race":

> Killing or the imperative to kill is acceptable only if it results not in a victory over political adversaries, but in the elimination of the biological threat to and the improvement of the species or race. . . . racism makes it possible to establish a relationship between my life and the death of the other that is not a military or warlike relationship of confrontation, but a biological-type relationship: "The more inferior species die out, the more abnormal individuals are eliminated, the fewer degenerates there will be in the species as a whole, and the more I—as species rather than individual—can live, the stronger I will be, the more vigorous I will be. I will be able to proliferate."[114]

Again, eugenics makes clear the impact of one's individual actions and life to the population as a whole. It allows the "I" to stand in for both the individual and the species, while at the same time delineating the difference between this "I" and the other. Negative eugenics—the sterilization or death of others—made this difference stark; eugenics, however, was not simply "negative eugenics" but also "positive eugenics"—the creation of a better species through positive choices in breeding. Indeed breeding encapsulates an early logic of programmability that inspired genetics and recognition of heredity as physical transmission. Eugenics, in other words, was not simply a factor driving the development of high-speed mass calculation at the level of content (the statistical demands of the biometricians helped foster mass calculation), but also at the level of logic or of operationality.

Breeding Programs

As Jann Sapp among others has claimed, genetics and eugenics were both intimately intertwined with the less elite practice of breeding—the *American Breeders' Magazine* (the official journal of the American Breeders' Association), for instance, had the subtitle *A Journal of Genetics and Eugenics*; breeders viewed heredity as an important economic force more wonderful than electricity since, once generated, it needed no additional force to sustain it.[115] Heredity was a living perpetual motion machine. François Jacob introduced the notion of a genetic program in *The Logic of Life* by arguing for breeding as an early "use" of heredity: "few phenomena in the living world are so immediately evident as the begetting of like by like . . . mankind early learnt to interpret and exploit the permanence of forms through successive generations. To cultivate plants, to breed animals, to improve them for food or to domesticate them, all require long experience."[116] Schrödinger, to repeat, used the steady transmission of certain features, such as the Hapsburg lip, as evidence of a genetic code-script. Breeding programs did not become separate, yet powerful, sources of power until after 1915, with the institutionalization of genetics.[117] The methodology of breeding—especially breeding mutant lines—was fundamental to the establishment of genetics as a science, that is, as a field that can produce hypotheses to be tested and predict future results regarding the transmission of discrete traits.

The interrelationship of eugenics, genetics, breeding, and capital was made most explicit by Charles Davenport, who was also a founder of the American Breeders' Association. According to Davenport, in *Heredity in Relation to Eugenics*, "eugenics is the science of the improvement of the human race by better breeding"; "human babies born each year constitute the world's most valuable crop"; and the goal of the eugenicist is to induce young people "to fall in love intelligently."[118] The references to breeding and other species were deliberate, and Davenport used them to emphasize the deficiencies of human reproductive control. "That marriage should be only an *experiment* in breeding, while the breeding of many animals and plants has been reduced to a science," he writes, "is ground for reproach. Surely the human product is superior to that of poultry; and as we may now predict with precision the characters of the offspring of a particular pair of pedigreed poultry so may it sometime be with man."[119] To produce such a science of heredity, one needed to delineate the Mendelian units responsible for—and the focus of—heredity. One needed to treat humans as carriers of definable genetic inheritances, which determined their worth and the cost of reproduction. Since direct human experimentation was not possible, Davenport and his helpers produced copious charts, based on human "history." Tracking everything from eye color to criminality, Davenport saw his work as essentially determining the independent unit characters and their impact on American society.

Mendelism hence made it much easier and more difficult to predict human heredity: easier because there were laws in place—most importantly, the law of dominance—but also harder because characters were not visible (but again, rather invisibly visible or visibly invisible). Early Mendelian genetics, because it relied on mechanistic laws rather than solely on statistics (in contrast to biometrics), could offer a strong notion of causal programmability: the notion that an invisible marker was responsible for a visible trait and that such a marker could be deleted through selective breeding programs. The goal of eugenics was to advise the government so that "good" blood could be fostered, especially with respect to immigration. The benefit of such breeding, as with domestic animals, was financial. Following the descendants of the legendary "Jukes" family of New York State, Davenport explains:

> Thus, in the same environment, the descendents of the illegitimate son of Ada are prevailingly *criminal*; the progeny of Bell are *sexually immoral*; the offspring of Ellie are *paupers*. The difference in the germ plasm determines the difference in the prevailing trait. But however varied the forms of non-social behavior of the progeny of the mother of the Jukes girls the result was calculated to cost the State of New York over a million and a quarter of dollars in 75 years—up to 1877, and their protoplasm has been multiplied and dispersed during the subsequent 34 years and is still marching on.[120]

Human genetics thus limited the effectiveness of government plans to "uplift" the nation through social welfare programs such as universal education or healthcare. "The expert teacher," Davenport claimed, "can do much with good material; but his work is closely limited by the protoplasmic makeup—the inherent traits—of his pupils."[121] So, while education was important, the best way to improve society was through better breeding:

> Indeed, while by good conditions we help the individual to make the most of himself, by good breeding we establish a permanent strain that is strong in its very constitution. The experience of animal and plant breeders who have been able by appropriate crosses to increase the vigor and productivity of their stock and crops should lead us to see that proper matings are the greatest means of permanently improving the human race—of saving it from imbecility, poverty, disease and immorality.[122]

Breedability became the proof of programmability in a bizarre logic that assumed any repetition evidence of inheritance, that is, repetition with no difference. It is programmability at its most rigorous. A self-propelled "living" repetition that also encapsulates death by condemning some to immediate death, and others to nonreproduction (thus ending repetition). This version of programmability also asserts a reverse-programmability, that is, the ability to determine an original algorithm—a strategy, or plan for action—based on interactions with unfolding events.

Programmability Continued

The fact that these claims could not be scientifically backed became clear as genetics began to examine the complexities of human heredity and of human populations. Many geneticists withdrew their early support of eugenics. By 1915, T. H. Morgan quietly severed his ties to the eugenics movement, although he did not speak publicly against the eugenics movement until many years later. According to Diane Paul, however, throughout the 1920s and 1930s, most geneticists remained supportive of eugenics, even though the scientific problems were well known by the 1920s.[123] As

Paul and William Provine among others have argued, the horrors of Nazi science forced many geneticists to review their relationship to eugenics and its assumptions and to speak publicly against it. According to Provine, however, the consensus that races did not vary hereditarily in intelligence did not take place until the 1950s.[124]

Although eugenics was eventually repudiated—and eugenics and genetics cannot be reduced to each other—the desire for the type of control openly embraced by eugenics did not fade. The confidence in and hope for a scientific future, Lily Kay has explicated in *The Molecular Vision of Life*, became folded into the project of molecular biology, which she reveals developed from the Rockefeller Foundation's "Science of Man" agenda. This agenda sought to "develop the human sciences as a comprehensive explanatory and applied framework of social control grounded in the natural, medical, and social sciences. Conceived during the late 1920s, the new agenda was articulated in terms of the contemporary technocratic discourse of human engineering, aiming toward an endpoint of restructuring human relations in congruence with the social framework of industrial capitalism."[125] As a form of human engineering, which cut across the various disciplines, it was a form of biopolitical governmentality. Even though the agenda changed in the 1930s, by which time eugenics had become a liability, Kay contends "the quest for rationalized human reproduction, however, never quite lost its intuitive appeal (even when it was later modified by the Nazi experience) . . . eugenic goals played a significant role in the conception and design of the molecular biology program." In particular, the failure of the old eugenics movement created a space for a new, more rigorous physiochemically based study of human heredity and behavior still focused on social betterment: "The molecular biology program, through the study of simple biological systems and analyses of protein structure," she writes, "promised a surer, albeit much slower, way toward social planning based on sounder principles of eugenic selection."[126]

The molecular vision of life, Kay relates, was amenable to strategies of control because it was governed by a faith in technology and in the ultimate power of upward causation.[127] It was control over nature through the study of the "ultimate littleness of things."[128] Molecular biology did not give up on mechanistic conceptions of life, but rather depended on the causal explanations of physics and chemistry.[129] The technological and the biological drove—and still drive—each other: "The enormous faith in the power of molecular genetics to explain human order and disorder has paralleled the enormous investments in genetic engineering in agriculture and medicine; the technological and cognitive realms drive and justify each other. This dialectic process of knowing and doing, empowered by a synergy of laboratory, boardroom, and federal lobby, has sustained the rise of molecular biology into the twenty-first century."[130] As Kay points out in her next book, *Who Wrote the Book of Life?*, this molecular vision of life would become supplemented by an "informational gaze," causing a profound rupture in "representations of life . . . from purely material and energetic to the informational."[131] As information displaced older visions of chemical and biological specificity and DNA was articulated as a programmed text, "the material control of life would be now supplemented by the promise of controlling its form and *logos*, its information."[132] In *Who Wrote the Book of Life?*, as noted earlier, Kay elegantly and convincingly reveals the ways in which information was adopted as a metaphor into biology in the period from 1953 to 1967.

This chapter has sought to show how this faith in and desire for a clear causality and programmability also dovetails with the rise of software as logos, as always already there, as something that persists and enables persistence. Agreeing that the move toward information changed discussions of biological control, it also argues that this turn to information as source did not simply emerge from elsewhere. Rather, it stemmed in part from the early eugenic belief in programmability, in an invisible mechanistic causality. The belief that DNA could be coded instructions thus was not simply a translation into biology of an idea already embedded within computer technology, but also something that preceded and indeed foreshadowed instructions becoming something in their own right. The bold belief in a code-script that could be both execution and plan found its real home in computers, which were constructed to enable this dream and then rediscovered as this dream come true. To be clear, this is not to say that computer programs are simply eugenic ones. Computer code-script conflates legislation with execution, but also reveals both the possibility and impossibility of a eugenic programmability: although "rules" may be followed, goals often are not. As well, code as logos resonates with the American eugenics movement's emphasis on individual decision making, rather than its overt message of social engineering. Important to this parallel structure—to this larger epistemic and governmental structure of programmability—is heredity as storage (this further turn of the helix is addressed in chapter 4). This dream of permanence, of something to be transferred in tact from generation to generation, makes lasting what is only ever ephemeral. The enduring ephemeral—that which repeats over and over again—becomes that which guarantees stability.

This chapter has also highlighted questions of neoliberal governmentality addressed in chapters 1 and 2. Whereas part I of the book focused on issues of hierarchy and bureaucracy, gender and labor, in part II, chapters 3 and 4 concentrate on how the computer, understood as code come true, encapsulates a certain logic of arranging, of taking care of, the relation between "men and things." It links the computer, and other projects of programmability, to biopower not simply in terms of content—the eugenics-based push for statistical analysis—but also in terms of logic. Cybernetics is a form of "governing," of navigating, in more than one sense. As Jacob notes, "the isomorphism of entropy and information establishes a link between two forms of power: the power to do and the power to direct what is done. . . . Information, an abstract entity, becomes the point of junction of the different types of order."[133] This juncture within neoliberalism becomes concentrated on the individual as human capital and individual decisions; governmentality is not simply or ever a process of overt engineering, of introducing death into a system focused on life and on freedom. In its neoliberal form, the form most resonant with computers, governmentality constructs the individual as both driven by and needing certain freedoms and desires that invisibly support a larger system. The term *liberal* contains within itself a reference to liberty. Foucault, elaborating on this necessary relationship between liberalism and liberty, claims that liberalism is "a consumer of freedom"

> inasmuch as it can only function insofar as a number of freedoms actually exist: freedom of the market, freedom to buy and sell, the free exercise of property rights, freedom of discussion, possible freedom of expression, and so on. . . . It consumes freedom, which means that it must produce it. It must produce it, it must organize it. The new art of government therefore appears as the management of freedom, not in the sense of the imperative: "be free," with the immediate contradiction that this imperative may contain. The formula of liberalism is not "be free." Liberalism formulates simply the following: I am going to produce what you need to be free. I am going to see to it that you are free to be free.[134]

Liberalism needs both to produce freedom and to devise mechanisms to control it. Neoliberalism, Foucault goes on to argue, focuses on the production and control of this freedom through competition. Thus, biopower as a form of power focused on "yes" rather than "no," on fostering life rather than death, encourages and is encouraged by a certain drive for life and independence, albeit one that is also linked to a tightly prescribed logic of programmability.

Both eugenics and software as logos have moved away from overt restrictions and toward a celebration of individual freedom and voluntary empowerment. Genetic counseling and birth control are not "negative eugenics"—the forced sterilization of the "weak"—but rather a framework for an informed decision that empowers those who engage it. Computing is not submission to the machine, but rather a means by which human intellect can be augmented (the race thus improved). This combination of empowerment with restrictions, of feelings of power generated by systems initially designed to restrict, drives the seductiveness of computing as a metaphor and as a way of encapsulating and experiencing power.

Relatedly, there have been movements in both biology and computation away from strict models of programmability, from the notion that the program or the map encapsulates the system or the organism. The rapidly growing discipline of systems biology—biology allegedly for the twenty-first century—for instance, emphasizes once more the importance of interactions or "communication" between cellular components, rather than reducing everything to the genetic code. Although it resonates with cybernetics and other early attempts to mathematically model biological systems, it stems from the "success" of the Human Genome Project, namely the deluge of data that project has produced, which has made clear the limitations of reductionism and traditional programmability (rather than producing algorithms that "explain" or encapsulate a behavior, programming is increasingly focused on using machine learning to "reverse engineer" the patterns driving data. Hence the term *data-driven programming*).

This turn to systems biology has also been driven by recent advances in mathematical models of biological systems, models that have been used to "validate hypotheses made from experimental data [and] designing and testing these models has led to testable experimental predictions. There are now impressive cases in which mathematical models have provided fresh insight into biological systems, by suggesting, for example, how connections between local interactions among system components relate to their wider biological effects."[135] This move away from specific genes and their corresponding functions (the idea that the genome is a code that can be cracked, that it is analogous to a software program that simply drives protein expression) toward a more nuanced understanding of the cell as comprised of various networks and signals (cell as an ecosystem) poses important new questions to science and technology studies, including a serious challenge to a politics that endorses the importance of nurture over nature. Considered for years as the automatically progressive, antiracist position, this new work on the importance of the environment—of local interactions—brings forward new, explicitly neoliberal questions of control. This is also clear in the move toward epigenetics. The widely cited article by Weaver and colleagues, "Epigenetic Programming by Maternal Behavoir,"[136] which showed that mice that were licked by their mothers were less anxious than their unlicked counterparts, as Hannah Landecker has argued, makes "good mothers" a focus of attention in problematic ways.[137] Catherine Malabou has similarly outlined the parallels between current neuronal understandings of the brain as a network and neoliberal management techniques, which emphasize "creativity, reactivity, and flexibility" and which also give the impression that "everyone . . . must take up the task of *choosing everything* and *deciding everything*."[138] Both brain as network and neoliberal management techniques move away from the notion of a central program or central power toward a decentralized network of agents.

Computing as well has moved toward less strictly "programmable" systems—in theory if not yet in everyday practice. From quantum, nonuniversal computers that lay down a path that can perhaps be taken to more software-based solutions such as genetic programs, the non-strictly programmable, the randomly produced is becoming ever more in vogue. The pressing question therefore is: What do we do with this move away from the map that nonetheless presupposes the map in a fundamental way? Another question then arises: What are both the drawbacks and the possibilities of becoming and processes, rather than of being and identity? Crucially, Malabou does not simply denunciate neurobiology, but rather engages it closely to argue for the difference between flexibility—which is capitulation—and plasticity. Plasticity, situated between two extremes—"the taking on of form" and "the annihilation of form"—enables a double movement, an explosive self-creation, that offers resistance to global neoliberalism.[139] How might we understand plasticity in relation to the ongoing transformation of programmable visions?

The Undead of Information

Computers have conflated memory with storage, the ephemeral with the enduring. Rather than storing memories, we now put things "into memory," both consciously and unconsciously. "Memory"—computer memory—has become surprisingly permanent. As Matthew Kirschenbaum has argued, our digital traces remain far longer than we suppose.[1] Hard drives fail, but can still be "read" by forensic experts (optically, if not mechanically); our ephemeral documents and other "ambient data" are written elsewhere—that is "saved"—constantly. Again, to read information is to write it elsewhere. At the same time, however, the enduring is also the ephemeral. Not only because even if data storage devices can be read forensically after they fail they still eventually fail, but also because—and more crucially—what is not constantly upgraded or "migrated" or both becomes unreadable. As well, our interactions with computers cannot be reduced to the traces we leave behind. The experiences of using—the exact paths of execution—are ephemeral. Information is "undead": neither alive nor dead, neither quite present nor absent.

Memory and storage are different. Memory stems from the same Sanskrit root for *martyr* and is related to the ancient Greek term for baneful, fastidious. Memory contains within it the act of repetition: it is an act of commemoration—a process of recollecting or remembering. In contrast, a store, according to the OED, stems from the Old French term *estorer* meaning "to build, establish, furnish." A store—like an archive—is both what is stored and its location. Stores look toward a future: we put something in storage in order to use it again; we buy things in stores in order to use them. By bringing memory and storage together, we bring together the past and the future; we also bring together the machinic and the biological into what we might call the archive.

Sigmund Freud famously modeled the human memory system, which he posited as fundamentally unconscious, on a toy called the *Mystic Writing Pad*. Describing the device, he wrote:

> The surface of the Mystic [Writing] Pad is clear of writing and once more capable of receiving impressions. But it is easy to discover that the permanent trace of what was written is retained upon the wax slab itself and is legible in suitable lights. Thus the Pad provides not only a

receptive surface that can be used over and over again, like a slate, but also permanent traces of what has been written like an ordinary paper pad . . . this is precisely the way in which, according to the hypothesis which I mentioned just now, our mental apparatus performs its perceptual function. The layer which receives the stimuli—the system *Pcpt.-Cs.* [Perception-Consciousness]—forms not permanent traces; the foundations of memory come about in other, adjoining, systems.[2]

According to Derrida, Freud, through this formulation posits a "prosthesis of the outside," which makes psychoanalysis a theory of the archive as well as of memory. It makes possible the "idea of an archive properly speaking, of a hypomnesic or technical archive, of a substrate or the subjectile (material or virtual) which, in what is already a psychic *spacing*, cannot be reduced to memory."[3] Memory in psychoanalysis is not first "live" and is not outside representation. Contemplating the importance of technology to this theory, Derrida asks, "Is the psychic apparatus *better represented* or is it *affected differently* by all the technical mechanisms for archivization and for reproduction . . . (microcomputing, electronization, computerization, etc.)?"[4] Intriguingly, the Mystic Writing Pad—or more properly its modern version, the Etch A Sketch®— returns as the model for the hard drive in a textbook on computer forensics. To explain the "unerasability" of hard drives, Warren G. Kruse II and Jay G. Heiser compare them to Etch A Sketches:

> When data is written onto magnetic media, a faint image of what was previously on the drive remains. A hard drive is like the child's drawing toy, the Etch A Sketch. Well, hard drives don't leak silver powder, but we are referring to the faint traces left after you erase an Etch A Sketch. The Etch a Sketch is erased by turning it over and shaking it, allowing the silver powder to coat the inside of the clear plastic window, preparing it for more drawings. But if you've used this popular toy, you'll remember that the faint traces of the previous drawing are always left behind. . . . Magnetic media—including hard drives—are similar in that every write leaves faint traces behind it, even when media have been overwritten numerous times.[5]

Data on a hard drive, Kruse and Heiser emphasize, leave a permanent trace, even as the drive makes room for new "impressions." This description of the hard drive, written by information security experts, eerily repeats Freud's description of the unconscious. It also highlights the work that "memory" (in contrast to archiving) entails—to be retrieved, these traces must be submitted to a rigorous process of reading.

How are we to understand archives as linking the machinic to the human to the written? As linking the ephemeral to the lasting? The alive to the dead? Two things to consider:

1. *The RNA world*   As mentioned previously, scientists are considering RNA more and more as primary. What is called the RNA world thesis argues that RNA is the "origin" of life, since RNA can act as both genes and enzymes and because DNA replication depends on "an enormous amount of proteins" (thus making DNA as origin unlikely).[6] Through retroviruses, RNA also rewrites DNA. This thesis fascinatingly questions the

conflation of legislation with execution that grounds code as logos. RNA does not simply code for proteins; DNA is no simple source.

2. *Cybernetics as memory* Jacques Derrida, in *Of Grammatology*, linked together writing and cybernetics: "The entire field covered by the cybernetic *program* will be the field of writing. If the theory of cybernetics is by itself to oust all metaphysical concepts—including the concepts of soul, of life, of value, of choice, of memory—which until recently served to separate the machine from man, it must conserve the notion of writing, trace, grammè [written mark], or grapheme, until its own historico-metaphysical character is also exposed."[7] Cybernetics, however, did not only have to conserve the notion of writing, but also that of memory. Memory links together the man and the machine. Memory also bridges across the machinic and human unknowns.

Moreover, to understand information as undead, we need to understand its relation to that other undead thing—the commodity. If a commodity is, as Marx famously argued, a "sensible supersensible thing," information would seem to be its complement: a supersensible sensible thing.[8] The literature, of course, on the relationship between information and the commodity is dense: from procapitalist celebrations of information as the new commodity to neo-Marxist ruminations on the impact of information on labor practices. Rather than rehearse these arguments, I want to emphasize that this parallel between information (as a general, rather than technical term) and commodities intersects with the emergence of source code as information outlined in chapter 1. That is, if information is a commodity, it is not simply due to historical circumstances or to structural changes; it is also because commodities, like information, depend on a ghostly abstraction.

Thomas Keenan, in "The Point Is to (Ex)Change It: Reading *Capital* Rhetorically," unpacks Marx's use of ghostly rhetoric to explain capital, in particular the capitalist exchange. Abstraction, Marx argues, transforms material things and their embedded-use values, into things that can be exchanged: commodities. This transformation fundamentally changes the "atomic" structure of things: "as exchange-values, [things] can be only different qualities, and thus not contain an atom . . . of use-value." Keenan asks: What, after this abstraction, is left? If exchange value eviscerates use—if it must eviscerate use to work—what makes possible exchange? What remains, Keenan contends, is a "ghost, *gespenstige Gegenständlichkeit*, spectral, haunting, surviving objectivity. 'There is nothing of them left over but this very same . . . ghostly objectivity, a mere jelly . . . of undifferentiated human labor.'" "This very phantom," Keenan goes on to insist, "makes possible the relation between (or within) things or uses, grants the common axis of similarity hitherto unavailable, precisely because it is a ghost and no longer a thing or a labor."[9] That ghostly jelly, Keenan argues, is humanity—the common humanity that survives in the things exchanged and, like language, makes exchange possible.

4 Always Already There, or Software as Memory

Software—as instructions and information (the difference between the two being erased by and in memory)—not only embodies the always already there, it also grounds it. It enables a logic of "permanence" that conflates memory with storage, the ephemeral with the enduring. Through a process of constant regeneration, of constant "reading," it creates an enduring ephemeral that promises to last forever, even as it marches toward obsolescence/stasis. The paradox: what does not change does not endure, yet change—progress (endless upgrades)—ensures that what endures will fade. Another paradox: digital media's memory operates by annihilating memory.

Remarkably, digital media has been heralded as "saving" analog media from destruction and obscurity. Many users, blind to the limitations of electromagnetic materials, assume that one can actually "store" things in memory. They assume that data saved on their DVDs, hard drives, and jump drives will always be there, that disk failure and the loss of memory it threatens are accidents instead of eventualities. Digitization surprisingly emerged as a preservation method in the 1990s by becoming a major form of "reformatting," a procedure designed to save intellectual content threatened by decaying materials—such as acidic wood-pulp paper and silver-nitrate film—by reproducing it.[1] Indeed, the National Endowment for the Humanities' 1988 "Brittle Books Program," which microfilmed millions of books in peril of "slow burn," viewed digitization as the preferred preservation method, even given a computer file's five-year shelf life. This celebration of the digital as archives' salvation stems in part from how digital files address another key archival issue: access. From the Library of Congress's early attempt to digitize its collections, the American Memory Pilot program (1990–1994), to Google's plan to digitize over ten million unique titles through its Book Search Program (announced in 2004), digitization has been trumpeted as a way for libraries finally to fulfill their mission: to accumulate and provide access to human knowledge. Digital archives are allegedly H. G. Wells's "World Brain" and André Malraux's museum without walls, among other dreams, come true.

At the same time, however, computer archives have been targeted as *the source* of archival decay and destruction, their liquidity threatening both the possibility and the

authenticity of cultural memory. Digital media disrupt the archive because they themselves are difficult to archive or have not been properly archived or both. The 1999
Modern Languages Association (MLA) report, "Preserving Research Collections: A Collaboration between Librarians and Scholars," summarizes the dual challenges of the
hard and the soft: "Imagine a historian opening a late nineteenth-century text and
helplessly watching as the title page breaks in her hand. Imagine another scholar, ten
years from now, inserting a disk containing an important document into the computer
and reading only a "fatal error" message on his screen. These two examples illustrate
the Janus-like preservation challenge faced by research libraries today: fragility of the
print past and the volatility of the future."[2] The material limits of materials not only
cause the future to be volatile, but also, again, so do the ever-updating, ever-prolifer-
ating, and increasingly incompatible soft and hard technologies—the challenges to
the historical preservation of software outlined in the introduction to this book.
Moreover, digital imaging potentially destabilizes authenticity. If libraries and archives,
as Abby Smith has argued, "serve not only to safeguard that information [which has
long-term value], but also to provide evidence of one type or another of the work's
provenance, which goes to establishing the authenticity of that work," this function
is seriously undermined by electronic images and documents, which are easily changed
or falsified.[3] The sheer plethora of digital files also calls into question the importance
of the libraries' and archives' traditional gatekeeping function. This is most clear in
the Internet Wayback Machine (IWM)'s approach to selection: this site creates a
"library of the Internet" by backing up all accessible sites. If libraries and archives
traditionally distinguished between materials of enduring value and "other bits of
recorded information, like laundry lists and tax returns," which were allowed to
vanish, the IWM has solved the extremely time-consuming task of selecting the enduring from the ephemeral by saving everything. (Although it originally tried to save
only "significant" material, it soon became an automatic archive of everything.) In
addition to all these difficulties, attempts to digitize content have been frustrated by
copyright issues, with rights holders demanding compensation or refusing permission.
Digital copies—allegedly defined by their immateriality—are, as the introduction has
emphasized, more closely regulated than their material counterparts, especially since
their use can be controlled by private contracts rather than by copyright or patents.

As this discussion makes clear, digital media's promise is also its threat; the two
cannot be neatly divided into the good and the bad. Digital media, if it "saves"
anything, does so by transforming storage into memory, by making what decays
slowly decay more quickly, by proliferating what it reads. By animating the inanimate—crossing the boundary between the live and the dead—digital media poses
new challenges and opportunities for "the archive."

Taking up the intertwining of the biological and the technological addressed previously, this chapter investigates how something as admittedly "soft" (and vapory) as

software hardened into something that allegedly guarantees heredity, and permanence. Looking in particular at von Neumann's early formulation of stored-memory computer architecture, chapter 4 argues that memory became conflated with storage through analogies to analogies: through analogies to cybernetic neurons, to genetic programs, to what would become "analog" media itself. Through these analogies (and their erasure), the new and the different have been reduced to the familiar. I uncover these differences and analogies not to attribute blame, but rather to reveal the dreams and hopes driving these misreadings: the desire to expunge volatility, obliterate ephemerality, and neutralize time itself, so that our computers can become synonymous with archives.[4] These desires are key to stabilizing hardware so that it can contain, regenerate, and thus reproduce what it "stores." Further, they are central to the twin emergence of neoliberalism and computer programs as strategic games.

These analogies also ground one of the fundamental axioms of digital media, namely that the digital reduces the analog—the real world—to 1s and 0s. By doing so the digital allegedly releases and circulates information that before clung stubbornly to material substances, effectively erasing the importance of context and embodiment. The fact that this has become an axiom should make us pause, especially since the evidence against it is substantial: the digital has proliferated, not erased, media types; what has become the analog is not the opposite, but rather the "ground" of the digital; and last, information is not naturally or inherently binary. Rather than making everything universally equivalent, the digital has exploded differences among media formats. Proprietary and nonproprietary electronic file formats such as jpeg, gif, mp3, QuickTime, doc, txt, rtf, and so on, not only distinguish between image, sound, and text, but also introduce ever more numerous differences among them. This explosion is not accidental to the digital, but rather, as I argue later, central to it. Also, the term *analog*, based on the word *analogy*, does not simply refer to what is real. After the emergence of electronic, arithmetically based computers, the term *analog* was adopted to describe computers that solved problems using similar physical models, rather than numerical methods. And finally, information is not simply digital, for information stems from the transmission of continuous electronic signals. The information traveling through computers is not 1s and 0s; beneath binary digits and logic lies a messy, noisy world of signals and interference. Information—if it exists—is always embodied, whether in a machine or an animal. To make information appear disembodied requires a lot of work, work that is glossed over if we just accept the digital as operating through 1s and 0s.

Revising the working thesis of chapters 2 and 3—software as axiomatic—chapter 4 contends that the digital is axiomatic. The digital emerges as a clean, precise logic through an analogy to an analogy, which posits the analog as real/continuous. Looking at the differences between analog and digital computers, this chapter reveals how discrete logical devices work by restricting possibilities and possible decodings.

It also examines how the development of these devices drives the need for "memory," a regenerating and degenerating archive that paradoxically, as Geoffrey C. Bowker notes, annihilates memory by substituting generalized patterns for particular memories.[5] This does not simply erase human agency, however, but rather fosters new dreams of human intervention, action, and incantation. It does not absolve us of responsibility, but instead calls on us to respond constantly, to save actively, if we are to save at all.

Biological Abstractions

John von Neumann's mythic, controversial, and incomplete 1945 "First Draft of a Report on the EDVAC" introduced the concept of stored program computing and memory to the U.S. military and the academic "public." This report is remarkably abstract: rather than describing actually existing components, such as vacuum tubes and mercury delay lines, it offers "hypothetical elements." According to von Neumann, it does so because, although dealing with real elements such as vacuum tubes would be ideal, such specificity would derail the process by introducing specific radio engineering questions at too early a stage. Thinking concretely in terms of types and sizes of vacuum tubes and other circuit elements "would produce an involved and opaque situation in which the preliminary orientation which we are now attempting would be hardly possible." To avoid this, von Neumann bases his consideration "on a hypothetical element, which functions essentially like a vacuum tube—e.g., like a triode with an appropriate associated RLC-circuit—but which can be discussed as an isolated entity, without going into detailed radio frequency electromagnetic considerations."[6] The vagaries of the machinery (vacuum tubes etc.), which are not necessarily digital but can be made to act digitally, threaten the clean schematic logic needed to design this clean, logical machine. Von Neumann describes this deferral as "only temporary."[7] However, J. Presper Eckert and John Mauchly, the original patent holders of stored program computing, would allege that von Neumann did not touch on the "true electromagnetic nature" of the devices because it was outside his purview: von Neumann, they contended, merely translated their concrete ideas into formal logic.[8] In fact, rather than a temporary omission, abstractness was von Neumann's modus operandi, central to the "axiomatic" (blackboxing) method of his general theory of natural and artificial automata and consonant with his game theory work.

This fateful abstraction, this erasure of the vicissitudes of electricity and magnetism, surprisingly depends on an analogy to the human nervous system. As cited earlier, von Neumann specifies the major components of the EDVAC as corresponding to different neurons: "The three specific parts CA [central arithmetic], CC [central control] (together C) and M [memory] correspond to the associative neurons in the human nervous system. It remains to discuss the equivalents of the *sensory* or *afferent* and the *motor* or *efferent* neurons. These are the *input* and the *output* organs of the device."[9] These neurons, however, are not simply borrowed from the human nervous system. They are the controversial, hypothetical neurons postulated by Warren McCulloch and Walter Pitts in their "A Logical Calculus of Ideas Immanent in Nervous Activity," a text McCulloch claims von Neumann saved from obscurity.[10] (Von Neumann would later describe these neurons as "extremely amputated, simplified, idealized.")[11] In accordance with McCulloch and Pitts, von Neumann expunges the messy materiality of these "neurons":

> Following W. S. McCulloch and W. Pitts . . . we ignore the more complicated aspects of neuron functioning: thresholds, temporal summation, relative inhibition, changes of the threshold by after-effects of stimulation beyond the synaptic delay, etc. It is, however, convenient to consider occasionally neurons with fixed thresholds 2 and 3, that is, neurons which can be excited only by (simultaneous) stimuli on 2 or 3 excitatory synapses (and none on an inhibitory synapse). . . . It is easily seen that these simplified neuron functions can be imitated by telegraph relays or by vacuum tubes. Although the nervous system is presumably asynchronous (for the synaptic delays), precise synaptic delays can be obtained by using synchronous setups.[12]

This analogy thus depends on and enables a reduction of both technological and biological components to blackboxes. In this simplified analogy, the effects of time are ignored to the extent that the synchronous can substitute for the asynchronous and interactions or "after effects" are erased.

So: to what extent are these abstractions and analogies necessary? What did and do they make possible? Clearly, this blackboxing, by divorcing symbolic analysis from material embodiment, has fostered a belief in information as immaterial, but more is at stake in this move to "biology." Notably, Claude Shannon's influential 1936 masters thesis, which showed that relay and switching can be symbolically analyzed (and designed) using Boolean logic, did not rely on an analogy between relays and neurons.[13] In *A Symbolic Analysis of Relay and Switching Circuits*, Shannon develops a means for simplifying and systematizing the development of complex electrical systems. He argues, "Any circuit is represented by a set of equations, the terms of the equations corresponding to the various relays and switches in the circuit." He then goes on to develop a calculus "for manipulating these equations by simple mathematical processes, most of which are similar to ordinary algebraic algorisms."[14] Shannon neither turns to biology nor elaborates on the material details of switches to ground his symbolic analysis. So why should the formal schematic of an automatic stored-memory computer be biologically inflected? And, why does a logical calculus—Boolean, digital logic—necessitate the erasure of the actual functioning of elements, such as vacuum tubes? To respond to these questions, I begin with another: How exactly are analog and digital related in electronic computing?

Nothing but Analog, All the Way Down

According to von Neumann in his 1948 "General and Logical Theory of Automata," a text that intriguingly reverses his initial analogy between vacuum tubes and neurons, the difference between "analogy and digital machines" lies in the ways they produce errors. Analogy machines, von Neumann contends, treat numbers as physical quantities. In order to perform a calculation, they thus find "various natural processes which act on these quantities in the desired way," such as wheel and disk integrators (which lie at the heart of the first computer mice). According to von Neumann, the guiding principle of analogy machines is the classic signal/information-to-noise ratio, a concept Shannon addresses in his *Mathematical Theory of Information*. That is, "the critical question with every analogy procedure is this: How large are the uncontrollable fluctuations of the mechanism that constitute the 'noise,' compared to the significant 'signals' that express the numbers on which the machine operates?"[15] If the calculation to be performed is complex and multistepped, such as the solving of partial differential equations, noise is amplified at every juncture, making it difficult to separate error from answer. Digital machines, in contrast, treat numbers as "aggregates of digits," rather than as physical quantities or signals. Because of this, they are not subject to noise constraints and offer the possibility of absolute precision, although von Neumann points out that round-off errors (now largely addressed by floating-point arithmetic) limit a digital machine's accuracy. Regardless, "the real importance of the digital procedure lies in its ability to reduce the computational noise level to an extent which is completely unobtainable by any other (analogy) procedure."[16]

Crucially, this reduction in noise occurs by ignoring the "analogy" aspect of digital components, for almost every element is a mixture of analogy and digit, as von Neumann acknowledges in "General and Logical Theory of Automata." In opposition to his "First Draft," this later article treats "living organisms as if they were purely digital automata." Responding to objections to this treatment, such as the fact that neurons do not simply work in an all-or-none fashion, he contends:

> In spite of the truth of these observations, it should be remembered that they may represent an improperly rigid critique of the concept of an all-or-none organ. The electromechanical relay, or the vacuum tube when properly used, are undoubtedly all-or-none organs. Indeed, they are the prototypes of such organs. Yet both of them are in reality complicated analogy mechanisms, which upon appropriately adjusted stimulation respond continuously, linearly or non-linearly, and exhibit the phenomena of "breakdown" or "all-or-none" response only under very particular conditions of operation.[17]

The digit, in other words, often treats a quantity as a discrete number, its accuracy resulting from a cut in a signal. The circularity of this passage, in which vacuum tubes are declared prototypes for all-or-none machines, is remarkable. Based on an analogy to computing elements, neurons, which themselves grounded computing elements as digital, are declared digital: an initial analogy is reversed and turned into ontology. At the base of this logic lies a redefinition of analogy itself as a complicated mechanism that operates on continuous quantities, rather than on discrete units.

This redefinition of analog as continuous, still present with us today whenever we refer to film and other media as "analog media," reveals a fundamental ambiguity at the core of what would become known as analog machines: does the analogy take place at the level of the machine architecture or at the level of signal? Analog as model emphasizes analogous differential equations and thus nonobvious analogous effects; analog as continuous buries these likenesses and privileges data over process. According to Thomas D. Truitt and A. E. Rogers in their 1960 *Basics of Analog Computers*:

> The word "analog" (or "analogue") has been used and misused. It has one meaning to some people, and a variety of uses to others. Webster speaks of a thing which maintains "a relation of likeness with another, consisting in the resemblance not of the things themselves, but of two or more attributes, or effects. . . . It is important to recognize that while *analog computer* refers most commonly to this one specific type of analog computer [general purpose d-c electronic analog computer], it can just as well refer to certain mechanical and hydraulic devices, to general purpose a-c electronic computers, and to a variety of special purpose computers. All of these have one characteristic in common—that the components of each computer or device are assembled to permit the computer to perform as a model, or in a manner analogous to some other physical system.[18]

Truitt and Rogers contend that similarities in system behavior, rather than resemblances between individual components, are key. In this sense, analog machines are simulation machines par excellence. Analog computers are based on similar physical relationships between mechanical and electronic systems and emphasize quantities over numbers. That is, the "signal" operated on and the result measured is a physical quantity, such as the intensity of an electrical current, or the rotation of a disk. Importantly, the notion of these machines as "analogy" machines only became apparent after the introduction of what would become digital computers, simulacra par excellence.

Analog to What?

Analog elements, even as they "ground" digital ones such as transistors and neurons, are not simple predecessors to digital computers. Analog and digital machines both thrived in the 1940s through the 1960s. Analog computers were used regularly in nuclear reactors for real-time data processing, as part of real-time control systems, such as flight simulators, and to simulate guided missiles in 3D (they were used to build the intercontinental ballistic missiles, which made the SAGE (Semi-Automatic Ground Environment) air defense system obsolete by the time it was completed).[19] So-called analog computers were popular because of their speed: they could solve problems in parallel, rather than serially (one step at a time), and although digital machines could complete one operation (such as subtraction) much more quickly than analog machines, they were not necessarily faster at complex operations. Early analog machines, as argued earlier, also offered a real-time graphical display that allowed engineers to see immediately how changing a coefficient or variable would alter a problem. Last, the fact that analog computers offered fewer decimal points in their solutions than their digital counterparts was often not important, since the accuracy of the calculation was frequently limited by other factors (measuring input, inadequate equations, etc.) and since early digital computers had significant digit control problems.

Not only were analog computers not viewed or accepted as stepping-stones toward digital ones, but also the division itself between analog and digital electronic computers was not clear. Electronic differential analyzers such as MADDIDA (Magnetic Drum Digital Differential Analyzer), which operated using Boolean algebra and digital electronic circuits, yet treated the signals to be operated as quantities rather than numerical entities, muddied the boundary between analog and digital machines —a boundary that arguably did not then exist. Indeed, analyzers only became analog computers rather than "mechanical mathematical" machines after electronics had displaced electromechanics in the production of discrete and nondiscrete machines.[20]

Electronics arguably marked a "break" between newer and older calculating machines in the 1940s as significant as the difference between digital and "analogy." In the May 1946 press release announcing the ENIAC (Electronic Numerical Integrator and Computer—the first working electronic digital computer), the U.S. War Department introduced it as the first "all-electronic general purpose computer," and underscored its "electronic methods."[21] Electronics marked the ENIAC's difference from both the mechanical "analog" differential analyzer and the "digital" (yet electromechanical) Harvard Automatic Sequence Calculator (Mark 1). In Vannevar Bush's 1945 Franklin Institute article introducing the electromechanical Rockefeller Differential Analyzer (RDA, built in 1942)[22] and in the press releases circulated that year, the RDA is never described by its makers/promoters as an analog machine, but rather as a "machine approach" to mathematics,[23] a "computing machine which marks a significant advance in the field of mechanized mathematics"[24] or, more colloquially, as an "electro-mechanical giant,"[25] a "tireless ally of science."[26] In response to these publications and to the War Department's announcing the ENIAC, newspapers reported on the machines together, calling them both "Magic Brains"[27] and "Mathematical Robots."[28]

Electronic devices were an important breakthrough because of their speed, and because they were built using nonspecialized labor. Mechanical differential analyzers required trained operators to be present at all times and inadvertently "taught" calculus to its "uneducated" operators. Bush claimed that the integraph (an early electronic version of the differential analyzer) enabled operators/students to cope with difficult mathematical questions by providing "the man who studies it a grasp of the innate meaning of the differential equation." For such a man, "one part at least of formal mathematics will become a live thing."[29] Seeing wheel and disk integrators in action makes calculus "live," moving it from formal writing to actual experience. According to Larry Owens, differential analyzers offered engineering students a graphic way to "think straight in the midst of complexity"—a type of thinking indebted to an engineering "graphical idiom," which operated as a universal language.

At the core of early analog analyzers lie ordinary differential equations. Similar ordinary differential questions describe seemingly disparate and unrelated electrical, electromechanical, mechanical, and chemical phenomena, all of which can be understood as closed "circuits." Analog machines, in this sense, work because ordinary differential equations are universal at a large scale, and because Newton's laws describing force can also describe electrical charge and water capacity.[30] For instance, the mechanical spring circuit represented in figure 4.1 corresponds to the RLC circuit in figure 4.2:

The mechanical spring system corresponds to the following formula:

$$m(d^2x/dt^2) = F \text{ [force]} - kx \text{ [oscillating force of spring]} - D(dx/dt)$$
$$\text{[dissipative force of friction]}$$

The electrical system of figure 4.2 has the following analogous differential equation (see table 4.1 for the corresponding quantities):

$$L(d^2q/dt^2) = V \text{ [voltage]} - 1/Cq \text{ [oscillating capacitor charge]} - R(dq/dt)$$
$$\text{[charge lost over resistor]}$$

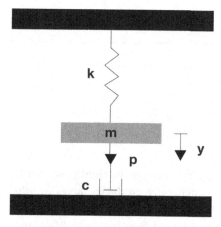

Figure 4.1
Mechanical spring circuit

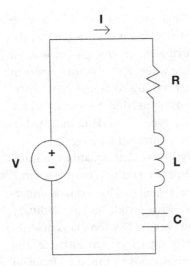

Figure 4.2
RLC circuit

Table 4.1
Analogous entities in the two systems

Mechanical	Electrical
force F	voltage V
mass m	inductance L
friction coefficient D	resistance R
displacement x	charge q
velocity dx/dt	current I
spring coefficient k	reciprocal of capacity $1/C$

All these equations could be put in the form

$D^{n-1}y/dx^{n-1} = \int d^n y/dx^n dx.$

For the mechanical spring system, this would be

$dx/dt \text{ [velocity]} = (1/m) \int (F - kx - D(dx/dt))dt.$

These equations are not usually solvable using normal analytic methods, but can be solved using numerical methods (desk calculators generally produced tables of solutions to differential equations before the popularization of machinic computers). MIT's differential analyzers employed a wheel and disc integrator to solve these differential

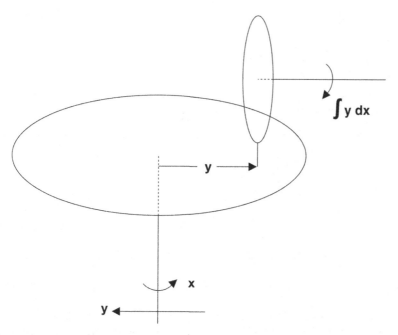

Figure 4.3
Schematic of a basic wheel and disk integrator

equations mechanically, using feedback to solve for values, which appeared on both sides of the equation sign. Figure 4.3 gives the basic design and principle of the integrator.

As figure 4.4 makes clear, the distance y is not a static value, but rather a function given determined by the rotation of another shaft.

So that

$W = k\int_{v1}^{v} U dV.$

To schematically represent the various operations, Bush used the following symbols (see figure 4.5):

So, using the equation $d^2y/dx^2 = f(x)$, in which case $f(x)$ is known in order to solve for y, one would build the setup outlined in figure 4.6.

Crucially, the differential analyzer employed "generative" functions—that is, the output could feed into itself. It could thus solve for variables on both sides of the equation. For instance, consider the solution for $d^2y/dx^2 = f(y)$, which is shown in figure 4.7.

These generative functions mark a fundamental difference between digital machines, which solve problems step by step, and analog machines.

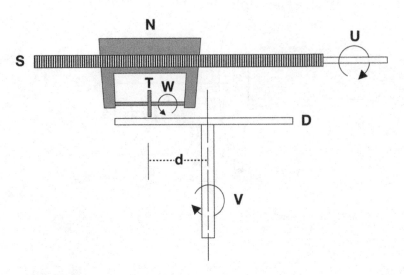

Figure 4.4
Integrator geometry

Because of this mechanical yet "live," analogous relationship, analog machines have generally been conceptualized as more transparent and intuitive than digital ones. Samuel Caldwell, director of MIT's Center of Analysis, stated, "There is a vividness and directness of meaning of the electrical and mechanical processes involved . . . [whereas] a Digital Electronic computer is bound to be a somewhat abstract affair, in which the actual computational processes are fairly deeply submerged."[31] Historian Paul Nyce has argued this mechanical mirroring made the move from analogy to essence or ontology difficult: one always dealt with—made visible—two analogous situations, rather than a universal solution. Nyce contends that analog devices

> belong to a long tradition of scientific instruments, starting in the seventeenth century, that "made visible what could not be seen" . . . Unlike most scientific instruments, however, analog devices supported both understanding (literally by measurement and number, like an astrolabe) and investigation for they, like an orrery, were "models" of phenomena. . . . What also made them persuasive is that they were both statements about and direct imitations of the things they represented. Mimesis is "hidden" or absent in digital machines: analog machines represent phenomena vividly and directly."[32]

Intriguingly, direct representation—or more accurately correspondence—makes analog machines live, vivid, and direct. It is a representation that always is tethered to another "source," which it does not try to hide. The differential analyzer was not, as the digital computer would be, amenable to notions of "universal" disembodied information. The differential analyzer simulated other phenomena, whereas digital computers, by

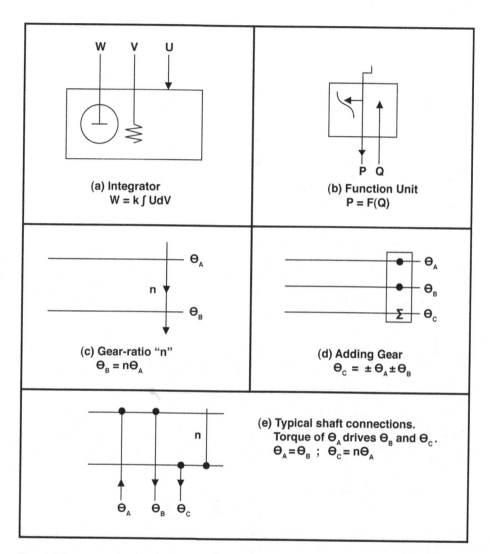

Figure 4.5
Symbols used in connection diagrams for a differential analyzer

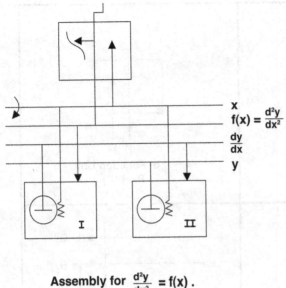

Assembly for $\dfrac{d^2y}{dx^2} = f(x)$.

Figure 4.6
$d^2y/dx^2 = f(x)$

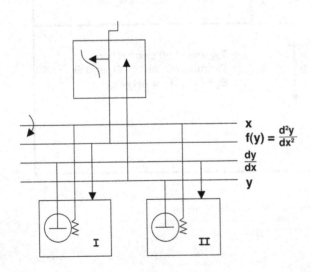

Assembly for $\dfrac{d^2y}{dx^2} = f(y)$.

Figure 4.7
$d^2y/dx^2 = f(y)$

hiding mimesis, could simulate any other machine. That is, while both digital and analog computers depend on analogy, digital computers, through their analogy to the human nervous system (which we will see stemmed from a prior analogy between neurons and Turing machines), simulate other computing machines using numerical methods, rather than recreating specific mechanical/physical situations. They move us from "artificial representation" or mechanical analysis (description) to simulacra or "information" (prescription). They move us from solving a problem by defining its parameters to solving it by laying out a procedure to be followed step by step. Depending on one's perspective, analog computers either offer a more direct, "intuitive," and, according to Vannevar Bush, "soul-satisfying" way of solving differential equations or they are imprecise and noisy devices, which add extra steps—the translation of real numbers into physical entities.[33] The first, the engineer's perspective, views computers as models and differential equations as approximations of real physical processes; the second, the mathematician's perspective, treats equations as predictors, rather than descriptors of physical systems—the computer becomes a simulacrum, rather than a simulation.

To be clear, though, analog machines did not simply operate via analogy; again, the notion that they operated through analogy would only be apparent later. They dealt with "signals," from which the notion and the theory of information would emerge, and further Vannevar Bush, as an electrical engineer, considered electricity to be a universal principle.[34] As well, to return to the question of electronics, all analog machines are not large, "intuitively understood," "live" mechanical devices. The electronic machines of the 1950s and 60s differed significantly from their mechanical predecessors. We thus need to be careful not to base arguments about analog machines as a whole on Vannevar Bush's early machines.[35] Indeed, Bush and Caldwell argued that one benefit of the electromechanical RDA was the fact that a trained operator was not necessary. As they explained, the user no longer had to "keep up "with the machine.[36] Op-amps as integrators, or even multipliers, were not "seeable" and graspable in the same fashion as wheel and disks.[37] Last, analog computational structures do not have to coincide perfectly with the problem to be solved: one can reuse an integrator in the same way that one can reuse an adder.

The move to electronics not only deskilled operators, it also made computers mass producible. The mechanical differential analyzers were steeped in the "odor" and the specialized labor of the machine lab, and they used special cams hand-crafted by highly skilled mechanics (the University of Pennsylvania Moore School Differential Analyzer was a WPA [Works Progress Administration] project, designed to employ mechanics). B. Holbrook, who worked at Bell Labs, argued that wire-wound potentiometers "offered the possibility of getting a completely new and relatively trainable type of labor into the manufacture of these things instead of the very high precision mechanics that were necessary by using the prior method."[38] Electronic analog and digital computers used mass-produced vacuum tubes and later transistors. Thus, both electronic analog and digital "machines" participate in Fordist logic: they automate calculation and production and make invisible the mathematics or calculations on which they rely.

Digital machines, however, are more profoundly Fordist than analog ones. The War Department ENIAC press release states that the ENIAC will eliminate expensive design processes: "Many electrical manufacturing firms, for instance, spend many thousands of dollars yearly in building 'analogy' circuits when designing equipment."[39] Most significant, they are more Fordist because their programming breaks down problems into simple, repeatable discrete steps. It is in programming, or to be more precise, programming in opposition to coding, that analog and digital machines most differ. Douglass Hartree, in his 1949 *Calculating Instruments and Machines*, reserves the terms *programming* and *coding* for digital machines, even though the RDA used tapes to specify the required interconnections between the various units, the values of ratios for the gearboxes, and the initial displacements of the integrators.[40] These tapes, unlike ones used for digital electronic computers, did not contain instructions necessary for sequencing a calculation; like von Neumann, Hartree describes programming as the "drawing up [of a] schedule of [the] sequence of individual operations required to carry out the calculation," and coding as the "process of translating operations into instructions in the particular form in which they are read by the machine."[41] Digital and analog electronic programming both retained the iconographic language of the differential analyzers, and in this sense were both grounded in mechanical methods or in their simulation. However, whereas digital flow charts produce a sequence of individual operations, analog programming produces a "circuit" diagram of systematic relations (see figure 4.8). These differences in programming also point toward key internal differences in representation, namely numbers versus quantities. Coded digital machines are much easier to follow. At a certain level then, analog machines (especially mechanical ones) were not simply more visual or transparent, but rather more complicated.

This complexity made it unlikely that analog computers could spawn or support code as logos. Code as logos—code as the machine—is intimately linked to digital design, which enables a strict step-by-step procedure that neatly translates time into space. Although later it would threaten to reduce all hardware to memory devices in the minds of most of its users, code as logos depended on a certain "hard" digital logic. This logic turns neurons and vacuum tubes themselves into logos and produces an insatiable need for memory, understood as regenerative circuits. This logic again stems from "biology," or, rather, from technologically enhanced biology: cybernetics.

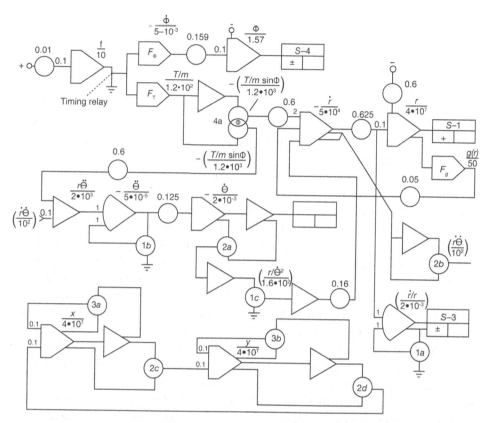

Scaled computer diagram for simulation of satellite system

Figure 4.8

An analog program diagram, based on an image from Albert S. Jackson, *Analog Computation* (New York: McGraw-Hill Book Company, 1960), 266

In the Beginning Was Logos (Again)

In "A Logical Calculus of the Ideas Immanent in Nervous Activity," McCulloch and Pitts seek to explain the operation of the brain in logical terms. This paper is part of McCulloch's larger project of "experimental epistemology," his effort to explain "how we know what we know . . . in terms of the physics and chemistry, the anatomy and physiology, of the biological system."[42] This experimental epistemology did not shun theory, but rather sought to weave together philosophy and neurophysiology. At its heart lies the equation of "the 'all-or-none' character of nervous activity" with propositional logic. It reduces a neuronal action to a statement capable of being true or false, "to a proposition which proposed its adequate stimulus."[43] This equation once more conflates word with action: in this particular case, the firing of a neuron with the proposition that "made" it fire. (Not surprisingly, McCulloch describes his examination of the human mind as a "quest of the Logos.")[44] This equation also concretizes the mind and ideas: "With the determination of the net," McCulloch and Pitts write, "the unknowable object of knowledge, the 'thing itself,' ceases to be unknowable."[45]

As the quotations around "all-or-none" imply, this description is a simplification, one coupled with assumptions such as: "a certain fixed number of synapses must be excited within a period of latent addition in order to excite a neuron at any time, and this number is independent of previous activity and position on the neuron."[46] Despite this, they argue that the all-or-none behavior of neurons makes them the fundamental psychic units or "psychons," which can be compounded "to produce the equivalents of more complicated propositions" in a causal manner.[47] Indeed, the goals of McCulloch and Pitts's logical calculus are to calculate the behavior of any neural net and to find a neural net that will behave in a specified way.[48] Remarkably, their method to "know the unknowable" not only simplifies nervous activity, it also does not engage the actual means by which inhibition or excitation occurs. This is because their method considers circuits equivalent if their result—their perceived behavior—is the same (as I explain later, this was crucial to cybernetic memory). Further, they erase actual alterations that occur during facilitation and extinction (antecedent activity temporarily alters responsiveness to subsequent stimulations of same part of the net) and learning (activities concurrent at some previous time alters the net permanently) via fictitious nets composed of ideal neurons whose connections and thresholds are unaltered.[49] Even though they state that formal equivalence does not equal factual explanation, they also insist that the differences between actual and idealized action do not affect the conclusions that follow from their formal treatment, namely the discovery/generation of a logical calculus of neurons.

Importantly, this logic of equivalence between neural nets and propositional logic was grounded, for McCulloch, in the nature of numbers themselves. In "What Is a Number, that Man May Know It, and a Man, that He May Know a Number?," he draws from David Hume to argue that only numbers truly can be equal. McCulloch's definition of numbers is Bertrand Russell's, "a number is the class of all those classes that can be put into one-to-one correspondence to it."[50] McCulloch's logical calculus, in other words, could only be digital with 1s and 0s corresponding to true and false. McCulloch later made this explicit, in his 1951 "Why the Mind Is in the Head," distinguishing the nervous system from sense organs in terms of digital versus analog. "In so-called logical, or digital contrivances," he writes, "a number to be represented is replaced by a number of things—as we may tally grain in a barn by dropping a pebble in a jug for each sheaf . . . the nervous system is par excellence a logical machine."[51] To McCulloch, logical equals digital because they both rely on numbers. Although analog machines also imply and are based on one-to-one models, McCulloch,

focusing on signals rather than on the machine, claims, "in so-called analogical contrivances a quantity of something, say a voltage or a distance, is replaced by a number
of whatnots or conversely, quantity replaces the number. Sense organs and effectors
are analogical."[52] In this schema, analog to digital conversion takes place at the level
of data—the difference in machine technology is completely erased through a logic
of equivalence.

By calling the cortex a digital machine, McCulloch sought to displace the then
popular theory of the mind as functioning mimetically. According to Seymour Papert,
McCulloch liberated the theory of perception from "the idea that there must be in the
brain some sort of genetically faithful representation of the outside world."[53] This is
most clearly seen in his 1959 "What the Frog's Eye Tells the Frog's Brain," (an article
with J. Y. Lettvin, H. R. Maturana, and W. H. Pitts). In it, they argue that because a
frog's eye does not transmit a copy of what it sees but rather detects certain patterns
of light and their changes in time, the "eye speaks to the brain in a language already
highly organized and interpreted, instead of transmitting some more or less accurate
copy."[54] Even earlier, though, and before von Neumann's preliminary draft, the cortex
for McCulloch was a Turing machine. In "A Logical Calculus," McCulloch and Pitts
state, "Every net, if furnished with a tape, scanners connected to afferents and suitable
efferents to perform necessary motor-operations, can compute only such numbers as
can a Turing machine."[55] Neural nets are inspired by and aspire to be Turing machines.[56]
Von Neumann's use of McCulloch and Pitts's analysis is thus an odd and circular way
of linking stored-memory digital computers to computing machines—once more, an
over-determined discovery of a linkage between biology and computer technology, yet
another turn of the double helix (before, of course, there was a double helix).

This linkage not only establishes a common formal logic, it also enables the emergence of computer "memory." Moving away from ideas of field-based, analogical
notions of memory, McCulloch's neural nets produce transitory memories and ideas
through circular loops. Drawing from Wiener's definition of information as order
(negative entropy), McCulloch argues that ideas are information: they are regularities
or invariants that conserve themselves as other things transform.[57] McCulloch conten-
tiously claims that this stability is produced by reverberating "positive-feedback"
circuits, that is, transitory memory (reverberatory memory cannot survive a "shut
down," such as a deep sleep or narcosis).[58] These reverberatory circuits, though, even
as they enable memory, also render "reference indefinite as to time past,"[59] for what
is retained is the memory, not all the events that led to that memory. In this sense,
they threaten to become "eternal ideas," separated from context. This separation,
combined with the fact that the neural nets can specify the next but not the previous
state, means that "our knowledge of the world, including ourselves, is incomplete as
to space and indefinite as to time."[60] Causality runs only one way: one cannot
decisively "reverse engineer" a neural net's prior state.

This emergence of memory is thus, as Bowker notes, also a destruction of memory. Thinking through cybernetician Ross Ashby's claim that "memory is a metaphor needed by a 'handicapped' observer who cannot see a complete system," Bowker writes, "The theme of the destruction of memory is a complex one. It is not that past knowledge is not needed; indeed, it most certainly is in order to make sense of current actions. However, a *conscious* holding of the past in mind was not needed: the actant under consideration—a dog, a person, a computer—had been made sufficiently different that, first, past knowledge was by definition retained and sorted and, second, only useful past knowledge survived."[61] What is truly remarkable is that this destruction of memory has spawned the seemingly insatiable need for computer memory. Memories are rendered into context-free circuits freed from memory, circuits that are necessary to the operation of the animal/machine.

Although the past may not be determinable from the present, memories—as context-free invariant patterns—ground our ability to predict the future. This prediction—causality—according to McCulloch (drawing from Hume) is only a "suspicion"[62] that there is "some law compelling the world to act hereafter as it did of yore."[63] Like those of ideas, these predictive circuits persist. Indeed, McCulloch argues, "the earmark of every predictive circuit is that if it has operated long uniformly it will persist in activity, or overshoot; otherwise it could not project regularities from the known past upon the unknown future."[64] The endurance of these circuits, however, threatens closure, threatens to make the unknown imperceptible, something that McCulloch "as a scientist . . . dread[s] most, for as our memories become stored, we become creatures of our yesterdays—mere has-beens in a changing world. This leaves no room for learning."[65] Memory, then, which enables a certain causality as well as an uncertainty as to time and place, threatens to overwhelm the system, creating networks that crowd out the new. A neural circuit, if it persists—programmability—makes prediction possible. It, however, also puts in jeopardy what for McCulloch is most interesting and vital about humanity: the ability to learn and adapt to the unknown, that is, the future as future.

This notion of memory as circuit/signal underscores McCulloch's difference from cognitive psychology, which, following developments in computer technology, would consider the brain hardware and the mind software.[66] In McCulloch's system, the mind and body are intimately intertwined, with the mind becoming less "ghostly"—more concrete—perhaps paradoxically by becoming signal.[67] Signals bridge mind and brain because they have a double nature; they are both physical events and symbolic values.[68] They are both statement and result. The logic of computers as logos stems from the disciplining, the axiomatizing, of hardware. This in turn "solidifies" instructions into things in and of themselves. Notably, McCulloch in his later work did address software, or programs, but referred to them as instructions to be operated on by data in memory, rather than as stored themselves in memory.[69] Instructions, in other words, did not drive the system—the logic, the logos, happened at the level of firing neurons.

Thus, by turning to McCulloch and Pitts rather than to Shannon, von Neumann gains a particular type of abstraction or logical calculus: an axiomatic abstraction and schematic design that greatly simplifies the behavior of its base components. Von Neumann also gains a parallel to the human nervous system, key to his later work on "general automata." Last, he "gains" the concept of memory—a concept that he would fundamentally alter by asserting the existence of biological organs not known to exist. Through this hypothetical "memory organ," and his discussion of the relationship between orders and data, his model would profoundly affect the development of cognitive science and artificial intelligence (AI) and life (AL). Through this memory organ, von Neumann would erase the difference between storage and memory, and also open up a different relationship between man and machine, one that would incorporate instructions—as a form of heredity—into the machine, making software fundamental. If word (as description) becomes event in McCulloch and Pitts's theory, in von Neumann's theory event once again becomes word, word becomes instruction.

Memories to Keep in Mind

Von Neumann's work with natural and artificial automata in general reverses the arrow of the analogy established in "First Draft." Rather than explaining computers in terms of the human nervous system, he elucidates the brain and its functioning in terms of computational processes. This is most clear in von Neumann's discussion of memory, which he considered to be a "much more critical and much more open" issue than logical processing.[70] In computer systems, memory was the bottleneck, for the limitations of memory on the machine created an "abnormal economy," in which the computer is forced to store all the information it needs to solve a problem on the equivalent of one page.[71]

The term *memory organ* clearly borrows from biology. This borrowing, however, was not necessary. Prior to "First Draft," mechanisms designed to store numbers and functions necessary for computing were called storage devices or "the store," following Babbage's terminology. J. Presper Eckert's 1944 "Disclosure of Magnetic Calculating Machine," used as evidence in the patent trial, refers concretely to the disks or tapes used to store data; his 1946 patent application, in contrast, employs the term *electrical memory*. This movement from storage to memory lies at the heart of the computer as archive, the computer as saving us from the past, from repetition through repetition.

Computer storage devices as memory is no simple metaphor, since it asserts the existence of an undiscovered biological organ. Although von Neumann initially viewed memory as comprising afferent neurons, he soon changed his mind, based on his own experience with computers, in particular with the number of vacuum tubes needed to create the types of reverberatory circuits McCulloch and Pitts described. In a reverse move, he postulated human memory as something unknown but logically necessary, making clear that his first analogy was based on a leap of faith. In *The Computer and the Brain*, written ten years after "First Draft," von Neumann writes, "the presence of a memory—or, not improbably, of several memories—within the nervous system is a matter of surmise and postulation, but one that all our experience with artificial automata suggests and confirms." Von Neumann goes on to emphasize our ignorance regarding this memory:

> It is just as well to admit right at the start that all physical assertions about the nature, embodiment, and location of [human memory] are equally hypothetical. We do not know where in the physically viewed nervous system a memory resides; we do not know whether it is a separate organ or a collection of specific parts of other already known organs, etc. It may well be residing in a system of specific nerves, which would then have to be a rather large system. It may well have something to do with the genetic mechanism of the body. We are as ignorant of its nature and position as were the Greeks, who suspected the location of the mind in the diaphragm. The only thing we know is that it must be a rather large-capacity memory, and that it is hard to see how a complicated automaton like the human nervous system could do without one.[72]

This passage reveals how quickly the computer moved from a system modeled on ideal neurons to a concrete model for more complex biological phenomena. This statement, which seems to be so careful and qualified—we basically do not know what the memory is or where it resides—at the same time asserts the existence of a memory organ or set of organs based on an analogy to computers: "The only thing we know is that it must be a rather large-capacity memory, and that it is hard to see how a complicated automaton like the human nervous system could do without one." This guess regarding capacity assumes that the brain functions digitally, that it stores information as bits, which are then processed by the brain, rather than functioning more continuously in a "field-based" manner. Again, this assumption was by no means accepted whole-heartedly by biologists. Dr. Lashley, among others, responded to von Neumann's difficulty with neuronal capacity by arguing that the memory was more dynamic rather than static and that "the memory trace is the capacity of many neurons to work together in certain permutations."[73]

Neurons as switching elements drive von Neumann's "logical" guess regarding memory capacity, as well as his confusion over its location:

> In the human organism, we know that the switching part is composed of nerve cells, and we know a certain amount about their functioning. As to the memory organs, we haven't the faintest idea where or what they are. We know that the memory requirements of the human organism are large, but on the basis of any experience that one has with the subject, it's not likely that the memory sits in the nervous system, and it's really very obscure what sort of thing it is.[74]

Digital switching devices, based on the reduction of all processes to true/false propositions, insatiably demand memoryless memory. As von Neumann explains in "First Draft," the need for memory increases as problems are broken down into long and complicated sequences of operations (described in chapter 1 of this book by Bartik and Holberton). Digital computation needs to store and have access to intermediate values, instructions, specific functions, initial conditions and boundary conditions, etc. Prior to the EDVAC, these were stored in an outside recording medium such as a stack of paper cards. The EDVAC was to increase the speed of calculation by putting some of those values inside the memory organ, making porous the boundaries of the machine. Memory instituted *a prosthesis of the inside.*[75] Memory was not simply sequestered in the "organ"; it also bled into the central arithmetic unit, which, like every unit in the system, needed to store numbers in order to work.

To contain or localize memory, von Neumann organized it hierarchically: there were to be many memory organs, defined by access time rather than content. For instance, in the 1946 work "Preliminary Discussion of the Logical Design of an Electronic Computing Instrument," von Neumann and colleagues divide memory into two conceptual forms—numbers and orders, which can be stored in the same organ if instructions are reduced numbers—and into two types—primary and secondary.[76] The primary memory consists of registers, made of flip-flops or trigger circuits, which need to be accessed quickly and ideally randomly. Primary memory, however, is very expensive and cumbersome. A secondary memory or storage medium supplements the first, holding values needed in blocks for a calculation. Besides being able to store information for periods of time, such a memory needs to be controllable automatically (without the help of a person), easily accessed by the machine, and preferably rewriteable. Interestingly, the devices listed as possible secondary memories are other forms of media: for instance, teletype tapes, magnetic wire or tapes, and movie film. (The primary media was also another medium—the Selectron was a vacuum tube similar to one used for television.)[77] This gives a new resonance to McLuhan's assertion that new media do not make preexisting media obsolete but merely change their use.[78] Von Neumann and colleagues also outlined a third form of memory, "dead storage," which is an extension of secondary memory, since it is not initially integrated with the machine. Not surprisingly, input and output devices eventually become part of "dead storage." As von Neumann argues later in *The Computer and the Brain*, "the very last stage of any memory hierarchy is necessarily the outside world, that is, the outside world as far as the machine is concerned, i.e. that part of it with which the machine can directly communicate, in other words the input and the output organs of the machine."[79] In this last step, the borders of the organism and the machine explode. Rather than memory comprising an image of the world in the mind, memory comprises the whole world itself as it becomes "dead."

This last step renders the world dead by conflating memory—which is traditionally and initially regenerative and degenerative—with other more stable forms of media such as paper storage, a comparison that is still with us today at the level of both memory (files) and interface (pages and documents). This conflation both relied on and extended neurophysiological notions of memory as a trace or inscription, like the grooves of a gramophone record. McCulloch, for instance, in 1951, in response to objections posed by von Neumann over memory as reverberatory circuits, outlined a hierarchical memory system that resonated with von Neumann's schema. There are first temporary reverberations, and second, nervous nets that alter with use (central to conditioned behaviors). The third type of memory, which he sees as an informational bottleneck, however, leaves him unhappily stumped; he is at a loss to describe its location and its operation:

> I don't see how we can tell where we have to look as yet, because in many of the experiments in which there are lesions made in brains, we have had large amounts of territory removed. However, usually we fail to destroy most fixed memories: therefore, we cannot today locate the filing cabinets. I think that sooner or later answers to the question of those filing cabinets, or whatever it is on which is printed "photographic records" and what not, will have to be found.[80]

The term *filing cabinet* is drawn from von Neumann's own terminology. In his response to McCulluch's paper, von Neumann, perhaps informed by psychoanalytical arguments that memories never die (one of von Neumann's uncles introduced psychoanalysis to Hungary and von Neumann apparently loved to analyze jokes) or by his personal experience (he allegedly had a photographic memory and could recall conversations word for word), presents the following "negative" and not entirely "cogent" argument against memory as residing in the neurons:

> There is a good deal of evidence that memory is static, unerasable, resulting from an irreversible change. (This is of course the very opposite of a "reverberating," dynamic, erasable memory.) Isn't there some physical evidence for this? If this is correct, then no memory, once acquired, can be truly forgotten. Once a memory-storage place is occupied, it is occupied forever, the memory capacity that it represents is lost; it will never be possible to store anything else there. What appears as forgetting is then not true forgetting, but merely the removal of that particular memory-storage region from a condition of rapid and easy availability to one of lower availability. It is not like the destruction of a system of files, but rather like the removal of a filing cabinet into the cellar. Indeed, this process in many cases seems to be reversible. Various situations may bring the "filing cabinet" up from the "cellar" and make it rapidly and easily available again.[81]

Von Neumann's "negative argument" relies on files and the human mind as the owner/manipulator—or, to return to Cornelia Vismann's argument outlined in chapter 2, chancellor—of files. It also depicts the human brain as surprisingly nonplastic: easily used up and unerased, hence once more the need for great storage. It also moves away from memory as based on erasable "regenerative" traces toward fantasies of traces that do not fade: immortality within the mortal machine.[82] This is a far cry from Vannevar Bush's description of the human mind in chapter 2 as fundamentally ephemeral and prone to forgetting. The digital paradoxically produces memory as storage, in part because logical algorithms need to read and write values. An entire process can fail if one variable is erased.

Memory as storage also allows von Neumann to describe genes as a form of human memory. In *The Computer and the Brain*, he writes, "another form of memory, which is obviously present, is the genetic part of the body; the chromosomes and their constituent genes are clearly memory elements which by their state affect, and to a certain extent determine, the functioning of the entire system."[83] With this move toward genes as memory—necessary for his theory of self-reproducing formula—neurons would not stand in for words (true or false propositions), but words (instructions) would come to stand in for neurons.

Descriptions that Can

The deed is everything, the Glory naught.
—*Faust*, Part II

According to William Poundstone, the last anecdote of von Neumann's "total recall" concerns his last days, when he lay dying of cancer at Walter Reed Hospital, a cancer caused by his work on nuclear weapons (the drive for nuclear weapons also powered the development of digital electronic computers; American computers and neoliberalism are both reactions to Nazism).[84] His brother Michael read *Faust* in the original German to von Neumann and, "as Michael would pause to turn the page, von Neumann would rattle off the next few lines from memory."[85] Converting to Catholicism before his death, von Neumann was deeply influenced by the work of Goethe, *Faust* in particular. Said his brother Nicholas, "We studied *Faust* in school very thoroughly, both parts, in original and in Hungarian translation. And we discussed it for years and rereading it occasionally thereafter, throughout our respective lifetimes."[86] One of the three passages Nicholas describes as particularly important to his brother was Faust's grappling with logos: "Faust's monologue at the opening of the First Part: 'In the beginning was the Act,' and the corresponding statement in Part II: 'The deed is everything, the Glory naught.' This we discussed in the context of the redeeming value of action."[87] According to Nicholas, this passage led "ultimately to John's views emphasizing the redeeming value of practical applications in his profession."[88] John von Neumann as an unredeemed (although not yet fallen) Faust.

This passage, however, has other resonances, intersecting with the question of logos weaving through this book. Faust, seeking to translate the Bible into German pauses over "in the beginning was the Word":

I'm stuck already! I must change that; how?

Is then "the word" so great and high a thing?

There is some other rendering,

Which with the spirit's guidance I must find.

We read: "In the beginning was the Mind."

Before you write this first phrase, think again;

Good sense eludes the overhasty pen.

Does "mind" set worlds on their creative course?

It means: "In the beginning was the Force."

So it should be—but as I write this too,

Some instinct warns me that it will not do.

The spirit speaks! I see how it must read,

And boldly write: "In the beginning was the Deed!"[89]

Faust, after a failed encounter with a spirit he conjured but cannot control, replaces Word with Deed, which, rather than Word, Force, or Mind, creates and rules the hour. Ironically, Faust, of course, is later saved by the Word—a technicality regarding his statement of satisfaction. Regardless, this substitution of Word with Deed sums up von Neumann's axiomatic approach to automata and his attraction to McCulloch and Pitts's work. It also leads him to conceive of memory as storage: as a full presence that does not fade, even though it can be misplaced. What is intriguing, again, is that this notion of a full presence stems from a bureaucratic metaphor: filing cabinets in the basement. This reconceptualization of human memory bizarrely offers immortality through "dead" storage: information as undead.

McCulloch and Pitts's methodology again depends on axiomatizing idealized neurons, where, according to von Neumann, "axiomatizing the behavior of the elements means this: We assume that the elements have certain well-defined, outside, functional characteristics; that is, they are to be treated as 'black boxes.' They are viewed as automatisms, the inner structure of which need not be disclosed, but which are assumed to react to certain unambiguously defined stimuli, by certain unambiguously defined responses."[90] This controversial axiomatization, which von Neumann would employ later in his theory of self-reproducing automata, reduces all neuronal activities to true/false statements.[91] Neurons follow a propositional logic. Von Neumann contends that this axiomatizing and the subsequent logical calculus it allows means that McCulloch and Pitts have proven that "any functioning . . . which can be defined at all logically, strictly, and unambiguously in a finite number of words can also be realized by such a formal neural network. . . . It proves that anything that can be exhaustively and unambiguously described, anything that can be completely and unambiguously put into words, is ipso facto realizable by a suitable finite neural network."[92] Words that describe objects, in other words, can be replaced by mechanisms that act, and all objects and concepts, according to von Neumann, can be placed in this chain of substitution. "There is no doubt," he asserts, "that any special phase of any conceivable form of behavior can be described 'completely and unambiguously' in words. This description may be lengthy, but it is always possible. To deny it would amount to adhering to a form of logical mysticism which is surely far from most of us."[93] This does not mean, however, that such a description is simple; indeed, von Neumann stresses that McCulloch and Pitts's theorizing is important for its reverse meaning: "there is a good deal in formal logics to indicate that the description of the functions of an automaton is simpler than the automaton itself, as long as the automaton is not very complicated, but that when you get to high complications, the actual object is simpler than the literary description."[94]

This notion of an actual object is not outside of language, even if it is outside "literary description," for, to von Neumann, producing an object and describing how to build it were equivalent. For instance, he argues that the best way to describe a visual analogy may be to describe the connections of the visual brain.[95] According to this logic, the instructions to construct a machine can substitute for the machine itself, to the extent that it can produce all the behaviors of the machine.

This logic is most clear in von Neumann's earliest model of self-reproduction, which Arthur Burks later dubbed a "robot" or "kinematic" model.[96] In this model, "constructing automata" A are placed in a "reservoir in which all elementary components in large numbers are floating."[97] Automaton A "when furnished the description of [an] other automaton in terms of appropriate functions will construct that entity." This description "will be called an instruction and denoted by a letter I. . . . All [As] have a place for an instruction I."[98] In this system, instruction drives construction. In addition to automata A, there are also automata B, which can copy any instruction I given to them. The decisive step, von Neumann argues, is the following instruction to the reader about embedding instructions:

> Combine the automata A and B with each other, and with a control mechanism C which does the following. Let A be furnished with an instruction I. . . . Then C will first cause A to construct the automaton, which is described by this instruction I. Next C will cause B to copy the instruction I referred to above, and insert the copy into the automaton referred to above, which has just been constructed by A. Finally, C will separate this construction from the system $A + B + C$ and "turn it loose" as an independent entity.[99]

This independent entity is to be called D. Von Neumann then argues, "In order to function, the aggregate $D = A + B + C$ must be furnished with an instruction I, as described above. This instruction, as pointed out above, has to be inserted into A. Now form an instruction I_D, which describes this automaton D, and insert I_D into A within D. Call the aggregate which now results E. E is clearly self-reproductive."[100] This instruction I_D (which nicely resonates with ID and id), he claims, is roughly equivalent to a gene. He also contends that B "performs the fundamental act of reproduction, the duplication of the genetic material, which is clearly the fundamental operation in the multiplication of living cells." This analogy fails, however, because "the natural gene does probably not contain a complete description of the object whose construction its presence stimulates. It probably contains only general pointers, general cues."[101] Thus, the memory of the system—here postulated as a more vibrant form of memory than "paper tape"—becomes the means by which the automaton can self-reproduce.[102]

This description is amazing for several reasons. In it, von Neumann transforms McCulloch and Pitts's schematic neural networks, in which there is no separation of software from hardware, into the basis for code as logos for the instructions replace the machine. What becomes crucial, in other words, and encapsulates the very being of the machine, are the instructions needed to construct it. Furthermore, and inseparable from the translation of event into instruction, this description—as a set of instructions itself—contains a bizarre, almost mystical, address. For, when von Neumann says, "Now form an instruction I_D, which describes this automaton D, and insert I_D into A within D," or "Combine the automata A and B with each other, and with a control mechanism C," who will do this forming and combining; who will perform these crucial steps and how? What mystical force will respond to this call? Like Faust before Mephistopheles arrives, are we to incant spells to create spirits? The transformation of description into instruction leaves open the question: who will do this? Who will create the magical description that goes inside? Remarkably, this call makes clear the fact that humans are indistinguishable from automata, something that bases von Neumann's game theory as well.

Games and Universes

This replacement of descriptions by instructions (or choices among instructions) also grounds von Neumann's work in game theory, which corresponds to his work on automata in many ways, as Arthur Burks has pointed out. "There is a striking parallel," Burks writes, "between von Neumann's proposed automata theory and his theory of games. Economic systems are natural competitive systems; games are artificial competitive systems. The theory of games contains the mathematics common to both kinds of competitive systems, just as automata theory contains the mathematics common to both natural and artificial automata."[103] This comparison, however, not only occurs at the level of mathematics or mathematization, but also at the level of heuristics, descriptions, and strategies. Game theory, which has been a key tool of neoliberal economic theory, seeks to understand the problem of exchange through the perspective of a "game of strategy," in which participants create strategies in response to others' moves, the rules of the game, and (objective) probabilities.[104] Similar to von Neumann's "First Draft," von Neumann and Oskar Morgenstern's 1944 *Theory of Games and Economic Behavior* (their preliminary discussion of game theory) serves as a *heuristic*, a "phase of transition from unmathematical plausibility considerations to the formal procedure of mathematics."[105] Also like his theory of automata, and indeed like most of von Neumann's mathematical work, game theory is based on an axiomatic method. Most importantly, von Neumann and Morgenstern introduce the notion of *strategy* to replace or simplify detailed description. Describing the process of giving an exact description of what comprises a game, they write, "we reach—in several successive steps—a rather complicated but exhaustive and mathematically precise scheme." Their key move is "to replace the general scheme by a vastly simpler one, which is nevertheless equivalent to it. Besides, the mathematical device which permits this simplification is also of an immediate significance for our problem: It is "the introduction of the exact concept of a strategy."[106] A strategy is a complete plan that "specifies what choices [the player] will make in every possible situation, for every possible actual information which he may possess at that moment in conformity with the pattern of information which the rules of the game provide for him for that case."[107] This replacement of a complete description with a strategy is not analogous to the replacement of machine code with a higher-level programming language, or what von Neumann calls "short code." This "equivalence" is not based on a simplification through the creation of a language that reduces several events into one statement, but rather on a fundamental transformation of a step-by-step description of events into a description of the premises—the rules and related choices—driving the player's actions. This strategy, which game theory remarkably assumes every player possesses before the game, is analogous to a program—a list of instructions to be followed based on various conditions. A player's strategy is not a summary of the rules of the game, but rather a list of choices to be followed—it is, to return to a distinction introduced in chapter 1, a product of "programming" rather than coding. Or, to put it slightly differently, understanding game strategy as a program highlights the fact that a program does not simply establish a universe as Weizenbaum argues; it is one possible strategy devised within an overarching structure of rules (a programming language). A strategy/program thus emphasizes the programming/economic agent as freely choosing between choices.[108]

   This program/strategy has been the basis of much of the criticism directed against game theory, such as Gregory Bateson's contention:

What applications of the theory of games do is to reinforce the player's acceptance of the rules and competitive premises, and therefore make it more and more difficult for the players to conceive that there might be other ways of meeting and dealing with each other. . . . Von Neumann's "players" differ profoundly from people and mammals in that those robots totally lack humor and are totally unable to "play" (in the sense in which the word is applied to kittens and puppies).[109]

Bateson is absolutely correct in his assessment: in outlining such a comprehensive version of a strategy, game theory assumes a player who could only be—or later would become—an automaton. Furthermore, von Neumann admits that game theory is prescriptive rather than descriptive. He writes, "the immediate concept of a solution is plausibly a set of rules for each participant which will tell him how to behave in every situation which may conceivably arise."[110] Thus, game theory presumes a strategy and the production of a strategy, as well as the replacement of a detailed description of every action with a more general procedural one. A strategy is something an automaton—or more properly a programmer—working non-"interactively" with a computer has. Game theory's assumptions again resonate with those of neoliberalism (Milton Friedman, to take one example, theorizes the day-to-day activities of people as analogous to those of "the participants in a game when they are playing it").[111]

Words, as instructions that stand in for deeds, are also crucial to von Neumann's desire to make his machines "universal." Von Neumann approaches the concept of universality through an interpretation of Alan Turing's "On Computable Numbers, with an Application to the Entscheidungsproblem," the 1936 paper that initially inspired McCulloch and Pitts.[112] In this paper, Turing shows that Hilbert's *Entscheidungsproblem* (the decision problem) cannot have a solution through theoretical machines, analogous to a "man," that can compute any number. He also posits the existence of a "universal machine," "a single machine which can be used to compute any computable sequence."[113] Von Neumann, in a rather historically dubious move, equates abstract or universal Turing machines with higher-level languages.

To make this argument, von Neumann separates codes into two types: complete and short. In computing machines, complete codes "are sets of orders, given with all necessary specifications. If the machine is to solve a specific problem by calculation, it will have to be controlled by a complete code in this sense. The use of a modern computing machine is based on the user's ability to develop and formulate the necessary complete codes for any given problem that the machine is supposed to solve."[114] Short codes, in contrast, are based on Turing's work, in particular his insight that "it is possible to develop code instruction systems for a computing machine which cause it to behave as if it were another, specified, computing machine."[115] Importantly, Turing himself did not refer to short or complete codes, but rather to instructions and tables to be mechanically—meaning faithfully—followed. Despite this, von Neumann argues that a code following Turing's schema must do the following:

> It must contain, in terms that the machine will understand (and purposively obey), instructions (further detailed parts of the code) that will cause the machine to examine every order it gets and determine whether this order has the structure appropriate to an order of the second

machine. It must then contain, in terms of the order system of the first machine, sufficient orders to make the machine cause the actions to be taken that the second machine would have taken under the influence of the order in question.

The important result of Turing's is that in this way the first machine can be caused to imitate the behavior of *any* other machine.[116]

Thus, in a remarkably circular route, von Neumann establishes the possibilities of source code as logos: as something iterable and universal. Word becomes action becomes word becomes the alpha and omega of computation.

Enduring Ephemeral

Crucially, memory is not a static but rather an active process. A memory must be held in order to keep it from moving or fading. Again, memory does not equal storage. Although one can conceivably store a memory, *storage* usually refers to something material or substantial, as well as to its physical location: a store is both what is stored and where it is stored. According to the OED, to store is to furnish, to build stock. Storage or stocks always look toward the future. In computer speak, one reverses common language, since one stores something in memory. This odd reversal and the conflation of memory and storage gloss over the impermanence and volatility of computer memory. Without this volatility, however, there would be no memory. To repeat, memory stems from the same Sanskrit root for *martyr*. Memory is an act of commemoration—a process of recollecting or remembering.

This commemoration, of course, entails both the permanent and the ephemeral. Memory is not separate from questions of representation or enduring traces. Memory, especially artificial memory, traditionally has brought together the permanent and the ephemeral; for instance, the wax tablet with erasable letters (the inspiration for classical mnemotechnics). As Frances A. Yates explains, the rhetorician treated architecture as a writing substrate onto which images, correlating to objects to be remembered, were inscribed. Summarizing the *Rhetorica Ad Herennium*, the classic Latin text on rhetoric, she states:

> The artificial memory is established from places and images . . . the stock definition to be forever repeated down the ages. A *locus* is a place easily grasped by the memory, such as a house, an intercolumnar space, a corner, an arch, or the like. Images are forms, marks or simulacra . . . of what we wish to remember. For instance, if we wish to recall the genus of a horse, of a lion, of an eagle, we must place their images on a definite *loci*.
>
> The art of memory is like an inner writing. Those who know the letters of the alphabet can write down what is dictated to them and read out what they have written. Likewise those who have learned mnemonics can set in places what they have heard and deliver it from memory. "For the places are very much like wax tablets or papyrus, the images like the letters, the arrangement and disposition of the images like the script, and the delivery is like the reading."[117]

Visiting these memorized places, one revives the fact to be recalled. This discussion of memory offers a different interpretation of the parallels between human and computer memory. The rhetorician was to recall a physical space within her mind—the image is not simply what is projected upon a physical space, but also the space for projection. Similarly, computer memory (which, too, is organized spatially) is a storage medium *like* but not quite paper. Both degenerate, revealing the limitations of the simile.

Memory as active process is seen quite concretely in early forms of "regenerative memory," from the mercury delay line to the Williams tube, the primary memory mentioned earlier. The serial mercury delay line (figure 4.9) took a series of electrical pulses and used a crystal to transform them into sound waves, which would make their way relatively slowly down the mercury tube. At the far end, the sound waves would be amplified and reshaped.[118] One tube could usually store about a thousand binary bits at any given moment.

Another early memory device, the Williams tube (figure 4.10), derived from developments in cathode ray tubes (CRTs); the television set is not just a computer screen, but was also once its memory. The Williams tube takes advantage of the fact that a beam of electrons hitting the phosphor surface of a CRT not only produces a spot of light, but also a charge. This charge will persist for about 0.2 seconds before it leaks

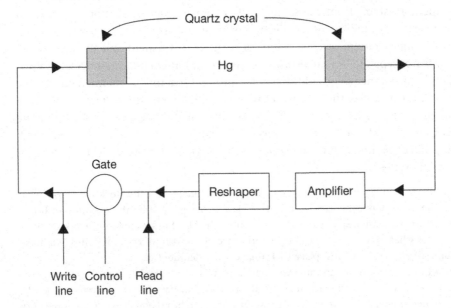

Figure 4.9
Schematic of the mercury delay line

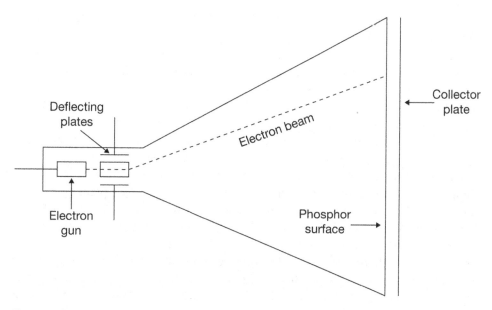

Figure 4.10
Schematic of the Williams tube

away and can be detected by a parallel collector plate. Thus, if this charged spot can be regenerated at least five times per second, memory can be produced in the same manner as the mercury delay tube. Current forms of computer memory also require regeneration.

Today's RAM is mostly volatile and based on flip-flop circuits and transistors and capacitors, which require a steady electrical current. Although we do have forms of nonvolatile memory, such as flash memory, made possible by better-insulated capacitors, they have a limited read-write cycle. Memory traces, to repeat Derrida's formulation, "produce the space of their inscription only by acceding to the period of their erasure."[119]

Thus, as Wolfgang Ernst has argued, digital media is truly *time-based media*, which, given a screen's refresh cycle and the dynamic flow of information in cyberspace, turns images, sounds, and text into a discrete moment in time. These images are frozen for human eyes only.[120] Information is dynamic, however, not only because it must move in space on the screen, but also, and more important, because it must move within the computer and because degeneration traditionally has made memory possible while simultaneously threatening it. Digital media, allegedly more permanent and durable than other media (film stock, paper, etc.), depends on a degeneration so actively denied and repressed. This degeneration, which engineers would like to divide into useful versus harmful (erasability versus signal decomposition, information versus noise), belies and buttresses the promise of digital computers as permanent memory machines. If our machines' memories are more permanent, if they enable a permanence that we seem to lack, it is because they are constantly refreshed—rewritten—so that their ephemerality endures, so that they may "store" the programs that seem to drive them. To be clear, this is not to say that information is fundamentally immaterial; as Matthew Kirschenbaum has shown in his insightful *Mechanisms: New Media and the Forensic Imagination*, information (stored to a hard drive) leaves a trace that can be forensically reconstructed, or again, as I've argued elsewhere, for a computer, to read is to write elsewhere.[121] This is to say that if memory is to approximate something so long lasting as storage, it can do so only through constant repetition, a repetition that, as Jacques Derrida notes, is indissociable from destruction (or in Bush's terminology, forgetting).[122]

This enduring ephemeral—a battle of diligence between the passing and the repetitive—also characterizes content today. Internet content may be available 24/7, but 24/7 on what day? Further, if things constantly disappear, they also reappear, often to the chagrin of those trying to erase data. When A3G (article III groupie), the gossipy conservative and supposedly female author of underneaththeirrobes.blogs.com—a blog devoted to Supreme Court personalities—came out as a thirty-year-old Newark-based U.S. attorney named David Lat in an interview with the *New Yorker*, his site was temporarily taken down by the U.S. government.[123] Archives of his site—and of every other site that does not reject robots—however, are available at the Internet Wayback Machine (IWM, web.archive.org) with a six-month delay.

Like search engines, the Internet Wayback Machine comprises a slew of robots and servers that automatically and diligently, and in human terms, obsessively, back up most web pages. Also like search engines, they collapse the difference between the Internet, whose breadth is unknowable, and their backups; however, unlike search engines, the IWM does not use the data it collects to render the Internet into a library, but rather use these backups to create what the creators of the IWM call a "library of the Internet." The library the IWM creates, though, certainly is odd, for it has no coherent shelving system: the IWM librarians do not offer a card catalog or a comprehensive, content-based index.[124] This is because the IWM's head librarian is a machine, only capable of accumulating differing texts. That is, its automatic power of discrimination only detects updates within a text. The IWM's greatest oddity, however, stems from its recursive nature: the IWM diligently archives itself, including its archives, within its archive.

The imperfect archives of the IWM are considered crucial to the ongoing relevance of libraries. The IWM's creators state: "Libraries exist to preserve society's cultural artifacts and to provide access to them. If libraries are to continue to foster education and scholarship in this era of digital technology, it's essential for them to extend those functions into the digital world."[125] The need for cultural memory drives the IWM and libraries more generally. Noting the loss of early film archives due to the recycling of early film stock, the archivists describe the imperative of building an "internet library":

> Without cultural artifacts, civilization has no memory and no mechanism to learn from its successes and failures. And paradoxically, with the explosion of the Internet, we live in what Danny Hillis has referred to as our "digital dark age."
>
> The Internet Archive is thus working to prevent the Internet—a new medium with major historical significance—and other "born-digital" materials from disappearing into the past. Collaborating with institutions including the Library of Congress and the Smithsonian, we are working to preserve a record for generations to come.[126]

The IWM is necessary because the Internet, which is in so many ways *about* memory, has, as Ernst argues, no memory—at least not without the intervention of something like the IWM.[127] Other media do not have a memory, but they do age and their degeneration is not linked to their regeneration. As well, this crisis is brought about because of this blinding belief in digital media as cultural memory. This belief, paradoxically, threatens to spread this lack of memory everywhere and plunge us negatively into a way-wayback machine: the so-called "digital dark age." The IWM thus fixes the Internet by offering us a "machine" that lets us control our movement between past and future by regenerating the Internet at a grand scale. The Internet Wayback Machine is appropriate in more ways than one: because web pages link to, rather than embed, images, which can be located anywhere, and because link locations always change, the IWM preserves only a skeleton of a page, filled with broken—rendered—links and images (figure 4.11). The IWM, that is, only backs up certain data types. These "saved"

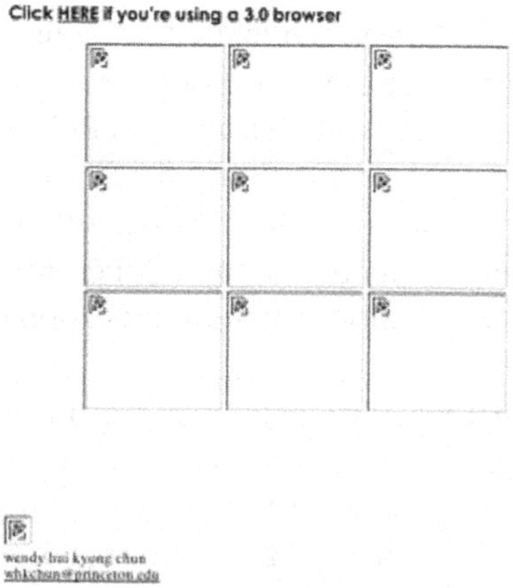

Figure 4.11

Screenshot of IWM backup of <http://www.princeton.edu/~whkchun/index.html>

pages are not quite dead, but not quite alive either, for their proper commemoration requires greater effort. These gaps not only visualize the fact that our constant regenerations affect what is regenerated, but also the fact that these gaps—the irreversibility of this causal programmable logic—are what open the World Wide Web as archive to a future that is not simply stored upgrades of the past.

Repetition and regeneration open the future by creating a nonsimultaneous new that confounds the chronological time these processes also enable. Consider, for instance, the temporality of weblogs (also known as *blogs*). Blogs seem to follow the timing of newspapers in their plodding chronology, but blogs contain within themselves archives of their posts, making the blog, if anything, like the epistolary novel. Unlike the epistolary novel, which, however banal, was focused on a plot or a moral, the blog entries are tied together solely by the presence of the so-called author. What makes a blog "uninteresting" is not necessarily its content, which often reads like a laundry list of things done or to do, but rather its immobility. The ever-updating, inhumanly clocked time in which our machines and memories are embedded and constantly refreshed makes the blog's material stale. The chronology, seemingly enabled by this time, is also compromised by these archives and the uncertainty of their regular reception. An older post can always be "discovered" as new; a new post is already old. This nonsimultaneousness of the new, this layering of chronologies, means that the gap between illocutionary and perlocutionary in high-speed telecommunications may be dwindling, but—because everything is endlessly repeated—response is demanded over and over again. The new is sustained by this constant demand to respond to what we do not yet know, by the goal of new media czars to continually create desire for what one has not yet experienced.

Digital media networks are not based on the regular obsolescence or disposability of information, but rather on the resuscibility or the undead of information. Even text messaging, which seems to be about the synchronous or the now, enables the endless circulation of forwarded messages, which are both new and old. Reliability is linked to deletion: a database is considered to be unreliable (to contain "dirty data"), if it does not adequately get rid of older, inaccurate information. Also, this repetition, rather than detracting from the message, often attests to its importance. Repetition becomes a way to measure scale in an almost inconceivably vast communications network.

Rather than getting caught up in speed then, what we must analyze, as we try to grasp a present that is always degenerating, are the ways in which ephemerality is made to endure. Paul Virilio's constant insistence on speed as distorting space-time and on real-time as rendering us susceptible to the dictatorship of speed has generated much good work in the field, but it can blind us to the ways in which images do not simply assault us at the speed of light.[128] Just because images flash up all of a sudden does not mean that response or responsibility is impossible, or that scholarly analysis is no longer relevant. As the news obsession with repetition reveals, an image does not flash up only once. The pressing questions are: why and how is it that the ephemeral endures? And what does the constant repetition and regeneration of information effect? What loops and what instabilities does the enduring ephemeral introduce to the logic of programmability? What is surprising is not that digital media fades, but rather that it stays at all and that we remain transfixed at our screens as its ephemerality endures.

Conclusion: In Medias Res

No matter how forewarned we are, thanks to the forearmaments of the knowledge of the secret of commodity exchange and its resulting fetishism, as long as exchange (language) goes on we are powerless to overcome its difficulties. And knowing makes it more scary. "Je sais bien, mais quand même." As Marx says, this is the path of madness: "If I state that coats or boots stand in a relation to linen because the former is the universal embodiment of abstract human labor, the craziness . . . of the expression hits you in the eye. But when the producers of coats and boots bring these commodities into relation with linen . . . the relation . . . appears to them in this crazy . . . form.". . . "Humanity" is this madness, its subject and its object. It is not simply the ignorance of not knowing what to do; it is rather the terror of still having to do, without knowing. And we have no magic caps, only ghosts and monsters.
—Thomas Keenan[1]

This book has traced the emergence of programmability through various theoretical and historical threads: code—both computer and genetic—as logos, user as sovereign, interfaces as "enlightening" maps, computer as metaphor for metaphor, and programmability as both thriving on and annihilating memory. It explores the extent to which computers, understood as networked software and hardware machines, are—or perhaps more precisely set the grounds for—neoliberal governmental technologies. And it examines how computers accomplish this not simply through the problems (population genetics, bioinformatics, nuclear weapons, state welfare, and climate) they make it possible to both pose and solve, but also through their very logos, their embodiment of logic.

The book began in part I with the question of code as logos, that is, with a "sourcery" that posited code written in higher-level programming languages as automatically and unfailingly "doing what it says." As the perfect performative utterance, code brought together two separate powers, the legislative and the executive, making execution and hardware largely irrelevant. This sourcery also opened part II, which posited genetics and computer code as complementary strands of a stylized double helix. Notably, code as logos within genetics precedes (rather than simply follows) its appearance in computer technology: Erwin Schrödinger famously hypothesized a genetic-code script that was "law-code and executive power . . . architect's plan and builder's craft—in one."[2] Both these cases, via their simplified maps of power, made possible the reemergence of a small-*s* sovereign subject, that is, one who could read and "speak"/manipulate these codes cum laws. This subject—described by Joseph Weizenbaum among others, as more powerful than emperors, playwrights, and chancellors—is the ultimate creator and master of completely knowable worlds. Not accidentally, both Alan Turing and Schrödinger, in explaining discrete computing and genetic codes respectively, return to "Laplace's view that from the complete state of the universe at one moment of time, as described by the positions and velocities of all particles, it should be possible to predict all future states."[3] Crucially, this power to know and create is not limited to programmers, but also spreads to users. Looking in particular at graphical user interfaces (GUIs) (which, along with higher-level programming languages, have been eroding the difference between users and programmers), this book outlines the ways in which computer interfaces "empower" users by amplifying their actions; this makes them the source of the action, putting them, in Douglas Engelbart's words, in the "bull-dozer's cab." Grounded on the principles of "direct manipulation" and "direct engagement," GUIs offer users a way to act and navigate an increasingly complex world. The maps they offer, as well as the paths they outline, seem to give individuals a way to comprehend their relationship to that "vaster and properly unrepresentable totality which is the ensemble of society's structures as a whole."[4] Fundamental to this mapping—which is offered by both computers and genetics—is memory, for memory makes possible both programs that link the past to the future and a "memoryless" inheritance. By storing programs and becoming archives, computers make the future predictable; by enabling a "hard" unconscious inheritance from generation to generation, genetics offers "you," as a form of human capital, powers (and "disabilities") that exceed personal experience. All of these threads together make the computer–genetics double helix a form of enlightenment: an empowering knowledge that finally enables man to forego his self-incurred tutelage and to be free. Freedom here stems from individual knowledge and actions, a central tenet of neoliberal governmentality.

At the same time, however, this book calls into question this narrative of overarching knowledge and freedom through these very concepts. Code as logos not only extends the power of individual programmers, it also makes code itself both legislation and execution: it spreads a neoliberal empowerment through the embedding of governmental enforcement into everyday situations, making us "subjects" of code. The gendered, military and eugenic histories examined bear witness to the constant anxiety that we programmers are not sovereigns, but slaves, that abstraction fosters greater ignorance. Further, the maps offered by GUIs are fundamentally mediated: as our interfaces become more "transparent" and visual, our machines also become more dense and obscure. Indeed, the call to map may be the most obscuring of all: by constantly drawing connections between data points, we sometimes forget that the map should be the beginning, rather than the end, of the analysis. Through our clicks, we perhaps always escape, but never leave, embroiled more strongly in an ideology that persists through our changes rather than our knowledge.[5] Our archive of knowledge as well seems to promise destruction and forgetfulness, as much as it promises permanence and stability. Memory would seem to be dynamic: an enduring ephemeral that disappears if it is not repeated (and also disappears through its repetition). So, instead of being enlightened and free, we seem to be caught in a certain madness: constantly acting without knowing, moving from crisis to crisis.[6] We seem to be free only within certain constraints, within a "mousetrap."

Crucially, this book has sought neither to condemn nor to celebrate software. Rather, it has been implicitly arguing that software can only be understood *in media res*—in the middle of things. In media res is a style of narrative that starts in the middle as the action unfolds. Rather than offering a smooth chronology, the past is introduced through flashbacks—interruptions of memory. To return to the parable of the six blind men relayed in the introduction, this means that the position of the blind men who know without knowing is not one to be superseded, but rather it is the position from which we can intervene and know. Software in media res also means that we can only begin with things—things that we grasp and touch without fully grasping, things that unfold in time, things that can only be rendered "sources" or objects (if they can) after the fact. Further, it means addressing the move within programming toward data-driven programming—a form of programming that, because it starts with data and then seeks (through machine learning algorithms) to discover the pattern "driving" behavior, is programming in media res. Last, in media res means taking seriously the computer's peculiar status as medium. It means grappling with the implications of the fact that a means of computation has also become a channel of communication and a storage device. Hence the emphasis on what is ghostly or undead, on what cuts across the human and the machine, on how we can make our interfaces more, rather than less, productively spectral; hence the emphasis on code as a *re-source*, rather than a source. Source code becomes a source only through its destruction, through its simultaneous nonpresence and presence.[7] Code (both biological and technological), in other words, is "undead" writing, a writing that—even when it repeats itself—is never simply a deadly or living repetition of the same.[8]

I thus want to end by addressing how being "in the middle of things," rather than in the driver's seat, can enable freedom and movement. My last book, *Control and Freedom: Power and Paranoia in the Age of Fiber Optics*, argued that freedom, rather than being the flip side of control, makes control possible, necessary, and never enough. In it I pondered freedom as an experience: as a giving over of ourselves to what is unknown, as an immersion that makes relation possible. Tellingly, although Milton

Friedman in his highly influential neoliberal theorization, *Capitalism and Freedom*, initially linked economic and political freedom, by 2002 (after a visit to Hong Kong), he delinked them, arguing "while economic freedom is a necessary condition for civil and political freedom, political freedom, desirable though it may be, is not a necessary condition for economic and civil freedom. . . . political freedom, which under some circumstances promotes economic and civic freedom, and under others, inhibits economic and civic freedom."[9] Political freedom cannot always support neoliberal economic freedom precisely because it is dangerous, because it is a dangerous experience that undoes, as much as it supports, the autonomous subject.

An anecdote relayed by Vicente Rafael in his insightful analysis of the 2001 People Power II protest makes this point nicely. This peaceful protest, which overthrew the government of Philippine President Joseph Estrada, in part was organized through text messages: illustrating how computer and networked technology can foster the desire for (and fear of) connection as much as they can for individualism. Neoliberalism is being superseded by neoconservatism: a drive to compensate for neoliberal isolation and chaos through a return to communal conservative values.[10] Neoconservatism, however, is not the only way in which networks are being reimagined. Rafael repeats the following online account of anonymous participant Flor C.'s encounter with the streaming crowd of people during the protest:

> When I first went to the flyover, I was caught in the thick waves of people far from the center of the rally. I could barely breathe from the weight of the bodies pressing on my back and sides. I started to regret going to this place that was [so packed] that not even a needle could have gone through the spaces between the bodies. After what seemed like an eternity of extremely small movements, slowly, slowly, there appeared a clearing before me (*lumuwag bigla sa harap ko*). I was grateful not because I survived but because I experienced the discipline and respect of one for the other of the people—there was no pushing, no insulting, everyone even helped each other, and a collective patience and giving way ruled (*kolektibong pasensiya at pagbibigayan ang umiral*).
>
> The night deepened. Hungry again. Legs and feet hurting. I bought squid balls and sat on the edge of the sidewalk. . . . While resting on the sidewalk, I felt such immense pleasure, safe from danger, free, happy in the middle of thousands and thousands of anonymous buddies.[11]

For Flor C., freedom stems from a collective patience and giving way—a collective flow in which one is immersed and imperiled. This freedom does not offer a feeling of mastery; it neither relies on maps nor sovereign subjects nor strategies, but rather depends on a neighborhood of relations and on unfolding actions. These actions, these movements reveal that "having to act without knowing" does not simply inspire terror—or if it does, it does not only do so; rather, such unknowing action makes possible collective human freedom.

Epilogue: In Medias Race

I want to conclude (again) by referring to a thread that has been largely invisible and yet central. This book initially was inspired by the striking parallels between software *in medias res* and race—that is, parallels between *software* and *race* as key terms in the current frenzy of and decline in visual knowledge. Linked together in the early twentieth century through the notion of a "genetic program," software and race embody two important ways of conceptualizing a seductively causal relationship between order and vision, the visible and invisible, the imaginable and readable—a causal relationship that contradicted early twentieth-century visions of a dark entropic future. Race and software are both nebulous entities (race cannot be scientifically defined; software cannot be physically separated from hardware), yet solid everyday experiences. We are expected to be as blind to software as we are to race; but race and software both act: both maintain visual literacy in an age of waning indexicality.

Like software, race was, and still is, a privileged way of understanding the relationship between the visible and invisible: it links visual cues to unseen forces. Interpreted through the lens of Mendelian genetics in the early- to mid-twentieth century, the consistent hereditability of racial features seemed to encapsulate an orderly transfer of traits, which belied the disorderly future predicted by statistical physics. A dream of order from order inspired conceptions of a strictly causal genetic code, which software—and not genetics—would be able to fulfill. Changes to conceptualizations of race are also key to understanding the vexed relationship between indexicality and causality this book has addressed. Although race since World War II no longer credibly links physical differences with innate mental differences, race remains a valid category. In the work of population geneticists, racial groups have become "breeding populations"; in the work of molecular biologists, racial groups are defined by the probability of having a combination of mainly unexpressed genetic material (the relationship between phenotype and genotype, which race supposedly explained, has resolved into DNA, and thus intersected with software). Culturally, as Toni Morrison has argued, race has become more on display than ever, even as the question of what race indexes—cultural or genetic differences, the results of economic injustice—remains unresolved.[1]

Race and software therefore mark the contours of our current understanding of visual knowledge as "programmed visions." As human vision is increasingly devalued through technological mediation in the sciences and through ideals of "color-blindness," images, graphics, and simulations proliferate. While writing this book, however, it became clear that the topic "race as archive" was too big to be included.[2] It has become a project in its own right, but I conclude with software in medias *race* because it has haunted this book and its vision.

You, Again

By now you should realize that there are many yous. Not simply because you adjoins the singular and the plural, but also because every you is haunted by what remains: by what remains as you read, by you as what remains.

The question: how are we to imagine you? By tracing the moments of connection—the ways in which the local unfolds to the global, constituting the "glocal"? Or, by taking these tracings as the beginnings of a more powerful imagining, a more powerful hallucination?

Notes

Preface

1. As I state in "Did Anyone Say New Media?," this demise does not coincide with the demise of media once called new, but rather with industry's quest to survive and thrive after the "new economy" and after new media's wide acceptance. Does it, after all, make sense to have a New Media Group within Apple in 2004? See *New Media, Old Media: A History and Theory Reader*, ed. Wendy Hui Kyong Chun and Thomas Keenan (New York: Routledge, 2006), 1–2.

Introduction

1. John Godfrey Saxe, "Blind Men and the Elephant," <http://www.wordinfo.info/words/index/info/view_unit/1/?letter=B&spage=3>, accessed 6/1/2008.

2. See Paul N. Edwards, *The Closed World: Computers and the Politics of Discourse in Cold War America* (Cambridge, Mass.: MIT Press, 1996).

3. The geneticists Luigi Luca and Francesco Cavalli-Sforza, for instance, argue: "It would be very difficult to change our hardware, our genetic makeup. It is much easier to try improving our software, our culture" in *The Great Human Diasporas: The History of Diversity and Evolution* (New York: Basic Books, 1995), 245.

4. Edwards, *Closed World*; Joseph Weizenbaum, *Computer Power and Human Reason: From Judgment to Calculation* (San Francisco: W. H. Freeman, 1976), 157.

5. Manfred Broy, "Software Engineering—From Auxiliary to Key Technology," in *Software Pioneers: Contributions to Software Engineering*, ed. Manfred Broy and Ernst Denert (Berlin: Springer, 2002), 11, 12.

6. Michael Mahoney, "The History of Computing in the History of Technology," *IEEE Annals of the History of Computing* 10, no. 2 (1988): 121.

7. As I argue later, however, all software is a reconstruction, a repetition.

8. Adrian Mackenzie, *Cutting Code: Software and Sociality* (New York: Peter Lang Pub. Inc., 2006), 169.

9. Herman H. Goldstine and John von Neumann, "Planning and Coding of Problems for an Electronic Computing Instrument," "Report on the Mathematical and Logical Aspects of an Electronic Computing Instrument," Part II, Volume I (Princeton, N.J.: Institute for Advanced Study, 1947), 2.

10. Paul Ceruzzi, *A History of Modern Computing*, 2nd ed. (Cambridge, Mass.: MIT Press, 2003), 80.

11. Friedrich Kittler, "There Is No Software," ctheory.net, October 18, 1995, <http://www.ctheory .net/text_file.asp?pick=74>, accessed 8/5/2010.

12. Mahoney, "History of Computing," 121.

13. Kathleen Broome Williams, *Grace Murray Hopper: Admiral of the Cyber Sea* (Annapolis, Md.: Naval Institute Press, 2004), 89; J. C. Chu, "Computer Development at Argonne National Laboratory," in *A History of Computing in the Twentieth Century*, ed. N. Metropolis et al. (New York: Academic Press, 1980), 345.

14. Martin Campbell-Kelly in *From Airline Reservations to Sonic the Hedgehog: A History of the Software Industry* (Cambridge, Mass.: MIT Press, 2003) divides the software industry into three period-based sectors: the software contractor (software as a service), software products, and mass-market products (3).

15. Herbert D. Benington, "Production of Large Computer Programs," *IEEE Annals of the History of Computing* 5, no. 4 (1983): 353.

16. This notion of programming is even evident in Fred Brooks's influential 1975 work *The Mythical Man-Month*, an analysis of the pitfalls of software programming based on his disastrous experiences with IBM's System/360. In it, he argues, "the purpose of a programming system is to make a computer easy to use"; in Frederick P. Brooks, *The Mythical Man-Month: Essays on Software Engineering*, 20th Anniversary Edition (New York: Addison-Wesley Professional, 1995), 43.

17. *Gottschalk v. Benson*, 409 U.S. 63 (1972); Pamela Samuelson, "Benson Revisited: The Case Against Patent Protection for Algorithms and Other Computer Program-Related Inventions," *Emory Law Journal* 39 (Fall 1990): 1053.

18. See Margaret Jane Radin, "Information Tangibility," in *Economics, Law, and Intellectual Property: Seeking Strategies for Research and Teaching in a Developing Field*, ed. Ove Granstrand (Dordrecht, The Netherlands: Kluwer Academic Publishers, 2003), 397.

19. *In re Alappat*, 33 F.3d at 1545, 31 USPQ2d at 1558 (Fed. Cir. 1994).

20. The U.S. Copyright Act of 1976 copyright states, "In no case does copyright protection for an original work of authorship extend to any idea, procedure, process, system, method of operation, concept, principle, or discovery, regardless of the form in which it is described, explained, illustrated, or embodied in such work."

21. Radin, "Information Tangibility," 407.

22. Rep. No. 473, 94th Cong., 1st Sess. 54 (1975).

23. 17 U.S.C. §102 (a).

24. Matthew Kirschenbaum, *Mechanisms: New Media and the Forensic Imagination* (Cambridge, Mass.: MIT Press, 2008).

25. Radin, "Information Tangibility," 406.

26. Ibid.

27. Ibid., 397.

28. She writes, "The compromise marries information to the external realm of objects. The structure of compromise involves the process of externalizing the internal ideas and embodying them in an object—a book. That is, the creative work starts out internal to the person, hence unpropertizable, but becomes embodied in an external object, hence propertizable" (Ibid., 398).

29. *The Oxford English Dictionary* (OED), 2nd ed., S.V. "information, *n.*"

30. Martin Heidegger, "The Thing," *Poetry, Language, Thought* (New York: HarperCollins, 1971), 176–177.

31. Colin Gordon, "Governmental Rationality: An Introduction," in *The Foucault Effect: Studies In Governmentality*, ed. Graham Burchell et al. (Chicago: Chicago University Press, 1991), 3. As Gordon explains, "Modern governmental rationalities consist, precisely, in the . . . 'daemonic' coupling of 'city-game' and 'shepherd-game': the invention of a form of secular political pastorate which couples 'individuation' and 'totalization'" (Ibid., 8). Foucault viewed sexuality, or, more broadly, technologies of the self, as crucial to the dual focus of both governmentality and his own research.

32. Foucault writes,

If I deploy the word "liberal," it is first of all because this governmental practice in the process of establishing itself is not satisfied with respecting this or that freedom, with guaranteeing this or that freedom. More profoundly, it is a consumer of freedom. It is a consumer of freedom inasmuch as it can only function insofar as a number of freedoms actually exist: freedom of the market, freedom to buy and sell, the free exercise of property rights, freedom of discussion, possible freedom of expression, and so on. The new governmental reason needs freedom, therefore the new art of government consumes freedom. It consumes freedom, which means that it must produce it. It must produce it, it must organize it. The new art of government therefore appears as the management of freedom, not in the sense of the imperative: "be free," with the immediate contradiction that this imperative may contain. The formula of liberalism is not "be free." Liberalism formulates simply the following: I am going to produce what you need to be free. I am going to see to it that you are free to be free. And so, if this liberalism is not so much the imperative of freedom as the management and organization of the conditions in which one can be free, it is clear that at the heart of this liberal practice is an always different and mobile problematic relationship between the production of freedom and that which in the production of freedom risks limiting and destroying it. . . . Liberalism as I understand it, the liberalism we can describe as the art of government formed in the eighteenth century, entails at its heart a productive / destructive relationship [with]* freedom. . . . Liberalism must produce freedom, but this very act entails the establishment of limitations, controls, forms of coercion, and obligations relying on threats etcetera. (Michel Foucault, *The Birth of Biopolitics: Lectures at the College de France, 1978–1979*, trans. Graham Burchell [Basingstoke, England, and New York: Palgrave Macmillan, 2008], 63–64)

Foucault's analysis of liberalism thus resonates strongly with my previous argument, in *Control and Freedom: Power and Paranoia in the Age of Fiber Optics* (Cambridge, Mass.: MIT Press, 2006), that computers have been intimately linked to the bizarre reduction of freedom to control and control technologies; but freedom, that book argued, cannot be reduced to control: freedom makes control possible, necessary, and never enough.

33. Foucault writes,

By this word [governmentality] I mean three things. First, by "governmentality" I understand the ensemble formed by the institutions, procedures, analyses and reflections, calculations and tactics that allow the exercise of this very specific, albeit very complex, power that has as its target population, political economy as its major form of knowledge, and apparatuses of security as it essential technical instrument. Second, by "governmentality," I understand the tendency, the line of force that, for a long time, and throughout the West, has constantly led towards the pre-eminence over all other types of power—sovereign, discipline, and so on—of the type of power which may be termed government, and which has led to the development of a series of specific governmental apparatuses . . . on the other hand . . . to the development of a series of knowledges. Finally, by "governmentality," I think we should understand the process, or rather, the result of the process, by which the state of justice of the Middle Ages became the administrative state during fifteenth and sixteenth centuries and was gradually "governmentalized." (Michel Foucault, *Security, Territory, Population: Lectures at the College de France, 1977–1978*, trans. Graham Burchell [Basingstoke, England, and New York: Palgrave Macmillan, 2007], 108–109)

34. For more on this, see Martin Campbell-Kelly and William Aspray, *Computer: A History of the Information Machine* (New York: Basic Books, 1996), 20–30.

35. Ibid., 22.

36. See David Alan Grier, *When Computers Were Human* (Princeton, N.J.: Princeton University Press, 2005).

37. David Harvey, *A Brief History of Neoliberalism* (Oxford: Oxford University Press, 2005), 2.

38. Foucault, *Birth of Biopolitics*, 117.

39. Friedman, *Capitalism and Freedom*, 4.

40. As quoted by David Harvey, *Brief History of Neoliberalism*, 23.

41. Foucault, *Birth of Biopolitics*, 252.

42. Ibid., 147.

43. Friedman, *Capitalism and Freedom*, 13; emphasis in original.

44. In this system, innovation "is nothing other than the income of . . . human capital, that is to say, of the set of investments we have made at the level of man himself" (Ibid., 231).

45. An intriguing symptom of this is the popularity of the *You* book series by Michael F. Roizen and Mehmet Oz. The first book is called *You: The Owner's Manual: An Insider's Guide to the Body that Will Make You Healthier and Younger* (New York: Harper, 2005).

46. Catherine Malabou, *What Should We Do with Our Brains*, trans. Sebastian Rand (New York: Fordham University Press, 2008), 44.

47. Ben Shneiderman, "Direct Manipulation: A Step Beyond Programming Languages," in *The New Media Reader*, ed. Noah Wardrip-Fruin and Nick Montfort (Cambridge, Mass.: MIT Press, 2003), 486.

48. As quoted in Fiona Morrow, "The Future Catches Up with Novelist William Gibson," *The Globe and Mail*, last updated April 3, 2009, <http://www.theglobeandmail.com/news/arts/article786109.ece>, accessed 9/9/2009.

49. See David Glovin and Christine Harper, "Goldman Trading-Code Investment Put at Risk by Theft," last updated July 6, 2009, *bloomberg.com*, <http://www.bloomberg.com/apps/news?pid=20601087&sid=a_6d.tyNe1KQ>, accessed 9/9/2009.

50. Joseph Weizenbaum has argued,

A large program is, to use an analogy of which Minsky is also fond, an intricately connected network of courts of law, that is, of subroutines, to which evidence is transmitted by other subroutines. These courts weigh (evaluate) the data given to them and then transmit their judgments to still other courts. The verdicts rendered by these courts may, indeed, often do, involve decisions about what court has "jurisdiction" over the intermediate results then being manipulated. The programmer thus cannot even know the path of decision-making within his own program, let alone what intermediate or final results it will produce. Program formulation is thus rather more like the creation of a bureaucracy than like the construction of a machine of the kind Lord Kelvin may have understood. (*Computer Power and Human Reason: From Judgment to Calculation* [San Francisco: W. H. Freeman, 1976], 234)

51. Bill Brown, "Thing Theory," *Critical Inquiry* 28, no. 1, Things (Autumn 2001): 4–5.

52. Ibid., 5.

53. Ibid., 3.

54. Ibid., 4.

55. Ibid., 5.

56. Ibid., 4.

57. Brown, "Thing Theory," 16; Heidegger, "The Thing," 165.

I Invisibly Visible, Visibly Invisible

1. As quoted by John Schwartz, "Privacy Fears Erode Support for a Network to Fight Crime," *New York Times*, March 15, 2004, C1. Seisint is a corporation developing The Matrix, a computer system described later in this section.

2. For more on this debate, see Thomas Elsaesser, "Early Film History and Multi-Media: An Archaeology of Possible Futures?," in *New Media, Old Media: A History and Theory Reader*, ed. Wendy Hui Kyong Chun and Thomas Keenan (New York: Routledge, 2006), 13–25; and Don E. Tomlinson, "One Technological Step Forward and Two Legal Steps Back: Digitalization and Television Newspictures as Evidence and as Libel," *Loyola of Los Angeles Entertainment Law Journal* 9 (1989): 237.

3. Roland Barthes, *Camera Lucida* (New York: Hill & Wang, 1981), 88.

4. Mary Ann Doane, *The Emergence of Cinematic Time: Modernity, Contingency, the Archive* (Cambridge, Mass., and London, England: Harvard University Press, 2002), 223.

5. *Crime Scene Investigation* (*CSI*) is a popular U.S. television program (also syndicated internationally), which features forensic investigations of crimes. Similarly, scientific digital animations of subcellular activity are the closest thing we have to true representations of activities that can be gleaned but not seen.

6. Jean Baudrillard. *The Ecstasy of Communication*, trans. Bernard and Caroline Schutze (Brooklyn, N.Y.: Autonomedia, 1988), 21–22; emphasis in original.

7. Information Awareness Office, Total Information Awareness Program (TIA) System Description Document, version 1.1 (July 19, 2002), <http://www.epic.org/privacy/profiling/tia/tiasystemdescription.pdf>, accessed 2/1/2007.

8. As quoted by Seth Finkelstein, "Google's Surveillance Is Taking Us Further Down the Road to Hell," *The Guardian* (March 26, 2009), <http://www.guardian.co.uk/technology/2009/mar/26/seth-finkelstein-google-advertising>, accessed 4/1/2009.

9. Joshua Gomez, Travis Pinnick, and Ashkan Soltani, "KnowPrivacy," June 1, 2009, <http://www.knowprivacy.org/report/KnowPrivacy_Final_Report.pdf>, 4, accessed 7/1/2009.

10. Paul Edwards, *The Closed World: Computers and the Politics of Discourse in Cold War America* (Cambridge, Mass.: MIT Press, 1996); Joseph Weizenbaum, *Computer Power and Human Reason: From Judgment to Calculation* (San Francisco: W. H. Freeman, 1976), 157.

11. Ibid.

12. Ibid., 234.

13. What, for instance, is Microsoft Word? Is it the encrypted executable—residing on a CD or one's hard drive—or its source code, or is it its execution? These, importantly, are not the same, even though they are all called Word: not only are they in different locations, but also two are programs while the other is a process. Structured programming, as discussed later, has been key to the conflation of program with process.

1 On Sourcery and Source Codes

1. Johann Wolfgang Goethe, "6. Faust's Study (i)," *Faust, Part One*, trans. David Luke (Oxford and New York: Oxford University Press, 1987), line 1236–1237.

2. As Barbara Johnson notes in her explanation of Jacques Derrida's critique of logocentrism, logos is the "image of perfectly self-present meaning . . . , the underlying ideal of Western culture. Derrida has termed this belief in the self-presentation of meaning, 'Logocentrism,' for the Greek word *Logos* (meaning speech, logic, reason, the Word of God)" in Translator's Introduction, *Dissemination*, trans. Barbara Johnson (Chicago: University of Chicago, 1981), ix.

3. For instance, *The A-2 Compiler System Operations Manual* prepared by Richard K. Ridgway and Margaret H. Harper under the direction of Grace M. Hopper (Philadelphia: Remington Rand, 1953) explains that a pseudo-code drives its compiler, just as "C-10 Code tells UNIVAC how to proceed. This pseudo-code is a new language which is much easier to learn and much shorter

and quicker to write. Logical errors are more easily found in information than in UNIVAC coding because of the smaller volume" (1).

4. Jacques Derrida, "Plato's Pharmacy," *Dissemination*, trans. Barbara Johnson (Chicago: University of Chicago, 1981), 115.

5. For example, George Landow, *Hypertext 2.0: The Convergence of Contemporary Literary Theory and Technology* (Baltimore, Md.: Johns Hopkins University Press, 1997); Sherry Turkle, *Life on the Screen: Identity in the Age of the Internet* (New York: Simon and Schuster, 1997); and Greg Ulmer, *Applied Grammatology: Post(e)-Pedagogy from Jacques Derrida to Joseph Beuys* (Baltimore, Md.: Johns Hopkins University Press, 1984). In her important work, N. Katherine Hayles has investigated the differences between poststructuralist writing and code in *My Mother Was a Computer: Digital Subjects and Literary Texts* (Chicago: University of Chicago Press, 2005).

6. Lev Manovich, *The Language of New Media* (Cambridge, Mass.: MIT Press, 2001), 48; emphasis in original.

7. *Vapor theory* is a term coined by Peter Lunenfeld and used by Geert Lovink to designate theory so removed from actual engagement with digital media that it treats fiction as fact. This term, however, can take on a more positive resonance, if one takes the nonmateriality of software seriously. See Geert Lovink, "Enemy of Nostalgia, Victim of the Present, Critic of the Future Interview with Peter Lunenfeld," July 31, 2000, <http://www.nettime.org/Lists-Archives/nettime-l-0008/msg00008.html>, accessed 2/1/2007.

8. Consider, for instance, the effectiveness of Chris Csikszentmihályi's largely unrealized *Afghan Explorer Robot*—covered by numerous news sources, it raised fundamental questions about restrictions on reporters covering war.

9. Alexander Galloway, *Protocol: How Power Exists after Decentralization* (Cambridge, Mass.: MIT Press, 2004), 164.

10. McKenzie Wark, "A Hacker Manifesto," version 4.0, <http://subsol.c3.hu/subsol_2/contributors0/warktext.html>, accessed 1/1/2010.

11. See Richard Stallman, "The Free Software Movement and the Future of Freedom; March 9th 2006," <http://fsfeurope.org/documents/rms-fs-2006-03-09.en.html>, accessed 6/1/2008. Immanuel Kant famously described the Enlightenment as "mankind's exit from its self-incurred immaturity" in "An Answer to the Question: What Is Enlightenment," in *What Is Enlightenment? Eighteenth-Century Answers and Twentieth-Century Questions*, ed. James Schmidt (Berkeley: University of California Press, 1996), 58.

12. For more on enlightenment as a stance of how not to be governed like that, see Michel Foucault, "What Is Critique?," in *What Is Enlightenment?*, ed. Schmidt, 382–398.

13. GNU copyleft/Public Licence (GPL) is symptomatic of the move in contemporary society away from the public/private dichotomy to that of open/closed. As Niva Elkin-Koren notes in "Creative Commons: A Skeptical View of a Worthy Project," the Creative Commons strategy "does not aim at creating a public domain, at least not in the strict legal sense of a regime that

is free of any exclusive proprietary rights. The strategy is entirely dependent upon a proprietary regime and drives its legal force from its existence. The normative framework assumes that it is possible to replace existing practices of producing and distributing informational works by relying on the existing proprietary regime. The underlying assumption is that if intellectual property rights remain the same, but rights are exercised differently by their owners, free culture would emerge" (<http://www.hewlett.org/NR/rdonlyres/6D4BFD1E-09BB-4F89-9208-7C1E4B141F2A/0/Creative_Commons_Amsterdam_final2006.pdf>, accessed 6/1/2008). Although Elkin-Koren is writing about Creative Commons in this passage, she makes it clear that this strategy of extending and revising intellectual property rights is drawn from the Free Software Movement's GPL.

14. Galloway, *Protocol*, 165–166; emphasis in original. Given that the adjective *executable* applies to anything that "can be executed, performed, or carried out" (the first example of *executable* given by the OED is from 1796), this is a strange statement.

15. See Derrida's analysis of *The Phaedrus* in "Plato's Pharmacy," 165–166.

16. John 1:1, *Bible, King James Version*.

17. Hayles, *My Mother Was a Computer*, 50. Hayles's argument immediately poses the question: What counts as internal versus external to the machine, especially given that, in John von Neumann's foundational description of stored program computing, the input and output (the outside world to the machine) was a form of memory?

18. Alexander Galloway, "Language Wants to Be Overlooked: Software and Ideology," *Journal of Visual Culture* 5, no. 3 (2006): 326.

19. Ibid.

20. Galloway, *Protocol*, 167.

21. Galloway, "Language Wants to Be Overlooked," 321.

22. This example draws from the *PowerPC Assembly Language Beginners Guide*, <http://www.lightsoft.co.uk/Fantasm/Beginners/Chapt1.html>, accessed 7/1/2009.

23. Paul Ricoeur, *The Rule of Metaphor*, trans. Robert Czerny et al. (London and New York: Routledge, 2003), 31.

24. *The Oxford English Dictionary* (OED), 2nd ed., S.V. "technical, *a.(n.)*"

25. Jacques Derrida stresses the disappearance of the origin that writing represents: "To repeat: the disappearance of the good-father-capital-sun is thus the precondition of discourse, taken this time as a moment and not as a principle of *generalized* writing. . . . The disappearance of truth as presence, the withdrawal of the present origin of presence, is the condition of all (manifestation of) truth. Nontruth is the truth. Nonpresence is presence. Differance, the disappearance of any originary presence, is *at once* the condition of possibility *and* the condition of impossibility of truth. At once" ("Plato's Pharmacy," 168).

26. Jacques Derrida, "Freud and the Scene of Writing," *Writing and Difference*, trans. Alan Bass (Chicago: University of Chicago Press, 1978), 228.

27. Herman H. Goldstine and John von Neumann, "Planning and Coding of Problems for an Electronic Computing Instrument," "Report on the Mathematical and Logical Aspects of an Electronic Computing Instrument," Part II, Volume I (Princeton, N.J.: Institute for Advanced Study, 1947), 2. They also state, "A coded order stands not simply for its present contents at its present location, but more fully for any succession or passages of C through it, in connection with any succession of modified contents to be found by C there, all of this being determined by other orders of the sequence."

28. Ibid.

29. Jacques Derrida in "Signature Event Context," in *Limited Inc.*, trans. Samuel Weber and Jeffrey Mehlman (Evanston, Ill.: Northwestern University Press, 1988 [1977]), argues, "every sign . . . can be *cited*, put between quotation marks; in so doing it can break with every given context, engendering an infinity of new contexts in a manner which is absolutely illimitable. This does not imply that the mark is valid outside of a context, but on the contrary that there are only contexts without any center or absolute anchoring [*ancrage*]. This citationality, duplication or duplicity, this iterability of the mark is neither an accident nor an anomaly, is that (normal/ abnormal) without which a mark could not even have a function called 'normal.' What would a mark be that could not be cited? Or one whose origins would not get lost along the way?" (12). Against this, Hayles, in *My Mother Was a Computer*, contends that computer code is not iterable because "the contexts are precisely determined by the level and nature of the code. Code may be rendered unintelligible if transported into a different context—for example, into a different programming language. . . . Only at the high level of object-oriented languages such as C++ does code recuperate the advantages of citability and iterability . . . and thus become 'grammatological'" (48).

30. For more on the software crisis and its relationship to software engineering, see Martin Campbell-Kelly and William Aspray, *Computer: A History of the Information Machine* (New York: Basic Books, 1996),196–203; Paul Ceruzzi, *A History of Modern Computing*, 2nd ed. (Cambridge, Mass.: MIT Press, 2003), 105; and Frederick P. Brooks's *The Mythical Man-Month: Essays on Software Engineering*, 20th Anniversary Edition (New York: Addison-Wesley Professional, 1995).

31. Philip E. Agre, *Computation and Human Experience* (Cambridge: Cambridge University Press, 1997), 92.

32. Norman Macrae, *John von Neumann* (New York: Pantheon Books, 1992), 378.

33. According to Friedrich Nietzsche, "there is no 'being' behind the doing, effecting, becoming: 'the doer' is merely a fiction added to the deed—the deed is everything." In *The Birth of Tragedy & The Genealogy of Morals*, trans. Francis Golffing (Garden City, N.Y.: Doubleday, 1956), 178–179.

34. Lawrence Lessig, *Code: And Other Laws of Cyberspace* (New York: Basic Books, 1999).

35. Code as automatically enabling and disabling actions is also code as police. I elaborate more on this in the "Crisis, Crisis, Crisis, or the Temporality of Networks" chapter of *Imagined Networks* (work in progress).

36. Milton Friedman, for instance, argues "the role of government just considered is to do something that the market cannot do for itself, namely, to determine, arbitrate, and enforce the rules of the game." See *Capitalism and Freedom*, Fortieth Anniversary Edition (Chicago: Chicago University Press, 2002), 27.

37. Ibid., 25.

38. Michel Foucault, *The Birth of Biopolitics: Lectures at the Collège de France, 1978–1979*, trans. Graham Burchell (Basingstoke, England, and New York: Palgrave Macmillan, 2008), 175.

39. David Golumbia, *The Cultural Logic of Computation* (Cambridge, Mass.: Harvard University Press, 2009), 224.

40. Judith Butler, *Excitable Speech: A Politics of the Performative* (New York: Routledge, 1997), 48.

41. Ibid., 78.

42. Ibid., 49; emphasis in original.

43. Joseph Weizenbaum, *Computer Power and Human Reason: From Judgment to Calculation* (San Francisco: W. H. Freeman, 1976), 234.

44. Martin Heidegger, "The Thing," *Poetry, Language, Thought* (New York: HarperCollins, 1971), 176–177.

45. Foucault, *Birth of Biopolitics*, 173.

46. None were classified as mathematicians, even though many of these women, hired to calculate ballistic trajectories using Marchant machines and even the differential analyzer, did have college degrees in mathematics (the "ENIAC girls," however, were promoted quickly to "mathematicians" as the war progressed). Special training courses in calculus and other "higher" maths were offered to those without college degrees by instructors such as Adele Goldstine, wife of Herman Goldstine and documenter of the ENIAC, as well as Mary Mauchly, wife of John Mauchly. This division between mathematicians and computers also existed during World War I, with graduate students and young assistant professors labeled as "computers," although the relations between them were less rigid. See David Grier, *When Computers Were Human* (Princeton, N.J.: Princeton University Press, 2005).

This notion of women as ideal programmers also persisted after the war with SAGE (discussed later) and projects such as Pacific Mutual's "Project Helpmate," a project that treated wives as interchangeable labor units to meet short-run staffing needs:

To do that, we had to make sure we didn't say that Ann is better than Sue and Sue is better than Jane. We employed them for the same salary amount, no matter what their experience, and for the same deal, which was that they would get so much a month and if they stayed til the end of the agreed upon time they got a bonus of so much. This was "project Helpmate." It was the only way we could get people, because we needed a lot of people for a limited period of time, and then wouldn't need them. Not only we wouldn't need them, but we thought we would need fewer of our other staff. That whole process is another story in itself. (Richard. D. Dotts, "Interview," May 21, 1973, *Computer Oral History Collection 1969–1973, 1977*, Archives Center, National Museum of American History, Smithsonian Institution [box 82], 23)

Women were also coders presumably because they would be more accepting of a job in which advancement was not clearly defined. As Dotts notes, programmers were viewed as easily expendible.

47. Jean J. Bartik, Frances E. Snyder Holberton, and Henry S. Tropp, "Interview," April 27, 1973, *Computer Oral History Collection, 1969–1973, 1977* (Washington, D.C.: Archives Center, National Museum of American History), <http://invention.smithsonian.org/downloads/fa_cohc_tr _bart730427.pdf>, accessed 8/8/2009, 12–13.

48. See "Women of the ENIAC: An Oral History," February 23, 2005 <http://www.witi.com/ center/aboutwiti/videos/eniac_kay_qt_hi.php>, accessed 8/8/2009.

49. Paul N. Edwards, *The Closed World: Computers and the Politics of Discourse in Cold War America* (Cambridge, Mass.: MIT Press, 1996), 71.

50. I. J. Good, "Pioneering Work on Computers at Bletchley," in *A History of Computing in the Twentieth Century: A Collection of Essays*, ed. N. Metropolis et al. (New York: Academic Press, 1980), 31–46.

51. Quoted in Ceruzzi, *A History of Modern Computing*, 82.

52. See Neal Stephenson, *In the Beginning . . . Was the Command Line* (New York: HarperCollins Publishers Inc., 1999).

53. See Heinz von Foerster, "Epistemology of Communication," in *The Myths of Information: Technology and Postindustrial Culture*, ed. Kathleen Woodward (Madison, Wisc.: Coda Press, 1980), 18–27.

54. See Norbert Wiener, "Cybernetics in History," *The Human Use of Human Beings* (New York: Da Capo Press, 1950), 15–27.

55. See Michael S. Mahoney, "Finding a History for Software Engineering," *IEEE Annals of the History of Computing* 26, no. 1 (January–March 2004): 8–19.

56. Johann Wolfgang Goethe, "7. Faust's Study (ii)," *Faust, Part One*, trans. David Luke (Oxford and New York: Oxford University Press, 1987), line 1648. The older translation by Philip Wayne uses the term *slave* (Baltimore, Md.: Penguin Books, 1949), 86.

57. According to Antonelli, they were hired after construction was well under way because it was feared that the assembled crew would disband soon (the machine was only operational after World War II).

58. Bartik, Holberton and Tropp, "Interview," 68.

59. To learn how to operate it, they went to Aberdeen to learn IBM punch card and plug board technology, and they studied the block diagrams fastidiously.

60. Bartik, Holberton and Tropp, "Interview," 54–55.

61. Ibid., 51.

62. Ibid., 38.

63. Ibid., 31.

64. Ibid., 33. Through this, she sought "to design something [so] that we could visualize what we were doing and also to devise something so that someone else could understand what we were doing to check it."

65. Ibid., 62. In essence, if you could take a motion picture of the machine, you would have been able to reproduce the numbers. The term *breakpoint* stems from the fact that the ENIAC programmers would actually pull the wire to stop the programs and read the accumulators (56).

66. Ibid., 56.

67. W. Barkley Fritz, "The Women of ENIAC," *IEEE Annals of the History of Computing* 18, no. 3 (September 1996): 21.

68. As argued earlier, although software produces visible effects, software itself cannot be seen. Luce Irigaray, in *This Sex Which is Not One*, trans. Catherine Porter (Ithaca, N.Y.: Cornell University Press, 1985), has similarly argued that feminine sexuality is nonvisual: "her sexual organ represents *the horror of nothing to see*" (emphasis in original, 26).

69. Sadie Plant, *Zeros + Ones: Digital Women + The New Technoculture* (New York: Doubleday, 1997), 37.

70. As quoted by Galloway, *Protocol*, 188.

71. Ibid.

72. Kathleen Broome Williams, *Grace Murray Hopper: Admiral of the Cyber Sea* (Annapolis, Md.: Naval Institute Press, 2004), 89; J. C. Chu, "Computer Development at Argonne National Laboratory," in *History of Computing*, ed. N. Metropolis et al., 345.

73. For example, see Fritz, "The Women of ENIAC," 13–28; Jennifer Light, "When Computers Were Women," *Technology and Culture* 40, no. 3 (July 1999): 455–483.

74. She nonetheless insisted that a woman who wanted to pursue mathematics in the Navy was a different story since "women have always done mathematics, since the days of the Greeks" (Hopper, "The Captain is a Lady," *60 Minutes Transcript*, March 1983, 11).

75. She states, "The novelty of inventing programs wears off and degenerates into the dull labor of writing and checking programs . . . this duty . . . looms as an imposition on the human brain." See "The Education of a Computer," *IEEE Annals of the History of Computing* 9, no. 3 (July–December 1987): 273. This rehumanizing of the mathematician was also to coincide with the mechanization (or perhaps, given the legacy of the ENIAC's "master programmer," the remechanization) of the professional computer.

Thus by considering the professional programmer (not the mathematician), as an integral part of the computer, it is evident that the memory of the programmer and all information and data to which he can refer is available to the computer subject only to translation into suitable language. And it is further evident that the computer is fully capable of remembering and acting upon any instructions once presented to it by the programmer.

With some specialized knowledge of more advanced topics, UNIVAC at present has a well-grounded mathematical education fully equivalent to that of a college sophomore, and it does not forget and does not make mistakes. It is hoped that its undergraduate course will be completed shortly and it will be accepted as a candidate for a graduate degree. (Ibid., 280–281)

The compiler was thus to enfold professional programmers within the machine itself, by dumping their memory into the machine.

76. Bartik, Holberton, and Tropp, "Interview," 105–106.

77. Goldstine and von Neumann, "Planning and Coding," section 7.9, 20, <http://www.admin .ias.edu/library/hs/da/ECP/Planning_Coding_Problems_v2p1.pdf>, accessed 8/1/2009.

78. Nathan Ensmenger and William Aspray, "Software as Labor Process," in *Proceedings of the International Conference on History of Computing: Software Issues* (New York: Springer-Verlag, 2000), 158.

79. Henry S. Tropp et al., "A Perspective on SAGE: Discussion," *IEEE Annals of the History of Computing* 5, no. 4 (October 1983): 387.

80. Martin Campbell-Kelly, *From Airline Reservations to Sonic the Hedgehog: A History of the Software Industry* (Cambridge, Mass.: MIT Press, 2004), 68–69. There was some resistance, as Irwin Greenwald points out, to this method: "We always separated the machine function from the procedures function or the programming function or what have you, which was very difficult for some people that we hired to get accustomed to. Stan Rothman, for example, was used to doing the whole job himself. I guess the years he was there he never did get to the point where he could accept our style. I know three and four different people who tried to work with him and gave up because he insisted on doing it the way he was used to, which was good, but not our style" (Greenwald and Robina Mapstone, "Interview," April 3, 1973, *Computer Oral History Collection 1969–1973, 1977, Archives Center, National Museum of American History, Smithsonian Institution,* Box 8, Folder 10, 17–18).

81. Philip Kraft, *Programmers and Managers: The Routinization of Computer Programming in the United States* (New York: Springer-Verlag, 1984), 15–16.

82. Tropp et al., "A Perspective on SAGE," 386.

83. Ibid., 385.

84. Ibid., 387.

85. Mahoney, "Finding a History," 15.

86. Kraft, *Programmers and Managers,* 99.

87. Edsger W. Dijkstra, "EWD 1308: What Led to 'Notes on Structure Programming,'" in *Software Pioneers: Contributions to Software Engineering,* ed. Manfred Broy and Ernst Denert (Berlin: Springer, 2002), 342.

88. Edsger W. Dijkstra, "Go To Statement Considered Harmful," in *Software Pioneers*, ed. Broy and Denert, 352.

89. Structured programming was introduced as a way to make the programs rather than the programmer priest the source, although the term *programmer priest* complicates the notion of *source:* is the source the programmer or some mythic power she mediates?

90. Edsger Dijkstra, "Notes on Structured Programming," April 1970, <http://www.cs.utexas.edu/~EWD/ewd02xx/EWD249.PDF>, accessed 8/8/2009, 7.

91. Ibid., 21.

92. John V. Guttag, "Abstract Data Types, Then and Now," in *Software Pioneers*, ed. Broy and Denert, 444.

93. Ibid.

94. Ibid., 445.

95. Thomas Keenan, "The Point Is to (Ex)Change It: Reading *Capital* Rhetorically," in *Fetishism as Cultural Discourse*, ed. Emily Apter and William Pietz (Ithaca, N.Y.: Cornell University Press, 1993), 165.

96. David Eck, *The Most Complex Machine: A Survey of Computers and Computing* (Natick, Mass.: A. K. Peters, 2000), 329, 238.

97. Adele Mildred Koss, "Programming on the Univac 1: A Woman's Account," *IEEE Annals of the History of Computing* 25, no. 1 (January–March 2003): 49.

98. Grier, *When Computers Were Human*.

99. Ibid., 33.

100. Ibid., 40.

101. After the Depression, it evolved into a more sophisticated unit, which used more complex numerical methods and Marchant machines to perform the basic mathematics.

102. Gertrude Blanch and Henry Tropp, "Interview," May 16, 1973, *Computer Oral History Collection 1969–1973, 1977, Archives Center, National Museum of American History, Smithsonian Institution,* <http://invention.smithsonian.org/downloads/fa_cohc_tr_blan730516.pdf>, accessed 8/8/2009, 3.

103. Ida Rhodes and Henry Tropp, "Interview," March 21, 1973, *Computer Oral History Collection 1969–1973, 1977, Archives Center, National Museum of American History, Smithsonian Institution,* <http://invention.smithsonian.org/downloads/fa_cohc_tr_rhod730321.pdf>, accessed 8/8/2009, 2.

104. Ibid.

105. Ibid., 3.

106. Blanch and Tropp, "Interview," 32.

107. Rhodes and Tropp, "Interview," 9.

108. "We were obsessed—especially Gertrude, who had fallen in love with the marvelous Tables of the British—with the idea that nothing but accuracy counts. And for this reason we had made it very clear to everybody that it was all [we] asked of them—complete accuracy" (Ibid., 12).

109. Ibid.

110. Blanch and Tropp, "Interview," 10.

111. Ibid., 12–13.

112. Alan Turing, "Proposal for Development in the Mathematics Division of an Automatic Computing Engine (ACE)," in *A. M. Turing's ACE Report of 1946 and Other Papers*, ed. B. E. Carpenter and R. W. Doran (Cambridge, Mass.: MIT Press, 1986), 38–39.

113. Koss, "Programming on the Univac 1," 56.

114. See Derrida, "Signature Event Context," 1–23.

115. *The A-2 Compiler System Operations Manual* prepared by Richard K. Ridgway and Margaret H. Harper under the direction of Grace M. Hopper (Philadelphia: Remington Rand, 1953), 53. This description makes it clear that pseudocode is not source. This manual, rather, refers to pseudo-code as information:

> The A-2 Compiler is a programming system for UNIVAC which produces, as its output, the complete coding necessary for the solution of a specific problem. If the problem has been correctly described to the compiler the coding will be correct and checked (by UNIVAC) and the program tape may be immediately run without any debugging. . . . The coding necessary for the solution of a specific problem is ordered by the programmer in pseudo-code. This pseudo-code is called "information" and it is the information which tells the compiler how to proceed just as C-10 Code tells UNIVAC how to proceed. This pseudo-code is a new language which is much easier to learn and much shorter and quicker to write. Logical errors are more easily found in information than in UNIVAC coding because of the smaller volume. (Ibid., 1; emphasis in original)

This passage nicely highlights the difference that pseudo-code makes: rather than dealing with the minutiae of machine programming (which in the case of UNIVAC was much easier than other machines because its native C-10 used alphanumerics rather than binary numbers), pseudo-code enables the programmer to address the logical problem at hand. Pseudocode was thus not conceived initially as "source code," but rather as an artifice—a supplement—that produced intermediate code that could be debugged.

116. Hardware was also an important limiting factor: the languages produced by the UNIVAC team did not penetrate widely because the UNIVAC did not; the Laining and Zierler system was limited to the Whirlwind.

117. John Backus, "Programming in America in the 1950s—Some Personal Impressions," in *History of Computing*, ed. N. Metropolis et al., 127.

118. Koss, "Programming on the Univac 1," 58.

119. Bob Everett, "Interview," May 29, 1970, *Computer Oral History Collection 1969–1973, 1977,* Archives Center, National Museum of American History, Smithsonian Institution (box 7, folder 2), 23.

120. Hopper goes on to say, "I should have studied a little harder but it wasn't that much that weighed you down and I just promptly relaxed into it like a featherbed and gained weight and had a perfectly heavenly time whereas the youngsters were very busy rebelling against the uniforms and the regulations and having to eat what was put in front of them and everything and I was just tickled to pieces by that time to have a beautiful meal put in front of me and I'd eat it." See Hopper and Tropp, "Interview," July 1, 1968, *Computer Oral History Collection 1969–1973, 1977,* Archives Center, National Museum of American History, Smithsonian Institution, <http://invention.smithsonian.org/downloads/fa_cohc_tr_hopp680700.pdf>, 26, accessed 9/9/2009. This freedom would make all the difference between her career trajectory (as a divorced woman with no children) and that of the other females around her who mostly left computing when they married or became pregnant.

121. United States Information Agency, *Voice of America Interviews with Eight American Women of Achievement* (College Park, Md.: National Archives and Records Administration, 1984), 9. Emblematically, Hopper loved drill, which she compared to dancing, and she became a battalion commander (Williams, *Grace Murray Hopper*, 22). The promotion to battalion commander also explains why she enjoyed the Navy so much. Midshipmen's school, after all, entailed both learning how to follow and how to give orders: "Thirty days to learn how to take orders, and thirty days to learn how to give orders, and you were a naval officer" (as quoted in Williams, *Grace Murray Hopper*). This learning to take and give orders made it possible for Hopper to reconcile discipline with leadership.

122. Relaying a visit to the Mark I, she notes, "nobody understood about computers, and they showed him how that they put the cards in the card reader and then they told this poor guy that those cards traveled down in that big round thing, and went up in the counter, and the poor guy believed it. They used to tell people the most horrible things, and I'm afraid we were at least partially responsible for some of the mythology that grew up around computers." See "Interview," *Computer Oral History Collection 1969–1973, 1977,* Archives Center, National Museum of American History, Smithsonian Institution, January 7, 1969, <http://invention.smithsonian.org/downloads/fa_cohc_tr_hopp690107.pdf>, accessed 9/9/2009, 12. For her "sufferings," see Williams, *Grace Murray Hopper*, 31. For her comment regarding experienced programmers, see G. M. Hopper and H. W. Mauchly, "Influence of Programming Techniques on the Design of Computers," *Proceedings of the IRE* 41, no. 10 (October 1953): 1250.

123. Harry Reed, "My Life with the ENIAC: A Worm's Eye View," in *Fifty Years of Army Computing: From ENIAC to MSRC*, ed. Thomas Bergin (U.S. Army Research Laboratory, 2000), 158; emphasis in original; accessible online at <http://ftp.arl.mil/~mike/comphist/harry_reed.pdf>.

124. Bartik, Holberton, and Tropp, "Interview," 121.

125. Ibid., 93.

126. Jean E. Sammet, *Programming Languages: History and Fundamentals* (Englewood Cliffs, N.J.: Prentice-Hall, 1969), 144.

127. Ibid., 148.

128. See, for instance, Theodor Nelson, *Computer Lib; Dream Machines* (Redmond, Wash.: Tempus Books of Microsoft Press, 1987).

129. David Golumbia's otherwise insightful analysis of computationalism also seeks to divide computing into two clear perspectives—the sovereign or the slave: "From the perspective we have been developing here, the computer encourages a Hobbesian conception of this political relation: one is either the person who makes and gives orders (the sovereign), or one follows orders. There is no room in this picture for exactly the kind of distributed sovereignty on which democracy itself would seem to be predicated" (*Cultural Logic of Computation*, 224).

130. See "Anecdotes: How Did You First Get into Computing?" *IEEE Annals of the History of Computing* 25, no. 4 (October–December 2003): 48–59.

131. Although not addressed here, there has always been tension between business and academic computing.

132. Frederick P. Brooks, while responding to the disaster that was OS/360, also emphasizes the magical powers of programming. Describing the joys of the craft, Brooks writes:

Why is programming fun? What delights may its practitioner expect as his reward?

First is the sheer joy of making things.

Second is the pleasure of making things that are useful to other people.

Third is the fascination of fashioning complex puzzle-like objects of interlocking moving parts and watching them work in subtle cycles, playing out the consequences of principles built in from the beginning.

Fourth is the joy of always learning, which springs from the nonrepeating nature of the task.

Finally there is the delight of working in such a tractable medium. The programmer, like the poet, works only slightly removed from thought-stuff. He builds his castles in the air, from air, creating by exertion of the imagination. . . . Yet the program construct, unlike the poet's words, is real in the sense that it moves and works, producing visible outputs separate from the construct itself. It prints results, draws pictures, produces sounds, moves arms. The magic of myth and legend has come true in our time. One types the correct incantation on a keyboard, and a display screen comes to life, showing things that never were nor could be. (Brooks, *Mythical Man-Month*, 7–8).

133. Paul Edwards, "The Army and the Microworld: Computers and the Politics of Gender Identity," *Signs* 16, no. 1 (Autumn 1990): 108–109.

134. Weizenbaum, *Computer Power and Human Reason*, 115; italics in original.

135. Foucault, *Birth of Biopolitics*, 120.

136. Indeed, he cites Horkheimer's critique:

Concepts have been reduced to summaries of the characteristics that several specimens have in common. By denoting similarity, concepts eliminate the bother of enumerating qualities and thus serve better to organize the material of knowledge. They are thought of as mere abbreviations of the items to which they refer. Any use transcending auxiliary, technical summarization of factual data has been eliminated as a last trace of superstition. Concepts have become "streamlined," rationalized, labor-saving devices . . . thinking itself [has] been reduced to the level of industrial processes . . . in short, made part and parcel of production." (Weizenbaum, *Computer Power and Human Reason*, 249)

And he argues that this critique applies directly to programming languages: "no one who does not know the technical basis of the systems we have been discussing can possibly appreciate what a chillingly accurate account of them this passage is. It was written by the philosopher-sociologist Max Horkheimer in 1947, years before the forces that were even then eclipsing reason, to use Horkheimer's own expression, came to be embodied literally in machines . . . As we see so clearly in the various systems under scrutiny, meaning has become entirely transformed into function" (Weizenbaum, *Computer Power and Human Reason*, 250).

137. Ibid., 255.

138. Brooks, *Mythical Man-Month*, 8.

139. Weizenbaum, *Computer Power and Human Reason*, 277.

140. Ibid., 115.

141. Ibid., 118.

142. Ibid., 124, 122, 121.

143. Ibid., 119.

144. For more on the relationship between paranoia and knowledge, rather than truth, see Wendy Hui Kyong Chun, *Control and Freedom: Power and Paranoia in the Age of Fiber Optics* (Cambridge, Mass.: MIT Press, 2006).

145. Ibid., 116.

146. Linus Torvalds and David Diamond, *Just for Fun: The Story of an Accidental Revolutionary* (New York: HarperCollins, 2001), 73.

147. *The Oxford English Online Dictionary* (Oxford and New York: Oxford University Press, 1992), <http://dictionary.oed.com/entrance.dtl>, accessed 3/1/2007.

148. See Friedrich Kittler, "There is No Software," *Ctheory* (1995), <http://www.ctheory.net/articles.aspx?id=74>, accessed 6/1/2008.

149. Namely, twentieth-century genetics, but this is the topic of chapter 3.

150. (Interview with) Felix Guattari and Gilles Deleuze, "Capitalism: A Very Special Delirium," in "Chaosophy," ed. Sylvere Lothringer, *Autonomedia/Semiotexte* (1995), <http://www.generation-online.org/p/fpdeleuze7.htm>, accessed 8/1/2009.

151. *The Oxford English Dictionary* (OED), 2nd ed., S.V. "fetish, *n.*"

152. William Pietz, "Fetishism and Materialism," in *Fetishism as Cultural Discourse*, ed. Apter and Pietz, 138, 139.

153. Ibid., 137.

154. Karl Marx, *Capital: A Critique of Political Economy, Volume One*, trans. Ben Fowkes (New York: Penguin Books in association with *New Left Review* 1990–1992, 1976), 165.

155. Marx as quoted by Pietz, "Fetishism and Materialism," 149.

156. Richard Stallman, "Copyright and Globalization in the Age of Computer Networks" (2001), <http://www.gnu.org/philosophy/copyright-and-globalization.html>, accessed 6/1/2008.

157. Ellen Ullman interviewed by Scott Rosenberg, "21st: Elegance and Entropy," *Salon.com*, <http://dir.salon.com/story/tech/feature/1997/10/09/interview/>, accessed 6/1/2008.

158. See Mez's site at <http://www.hotkey.net.au/~netwurker/>, accessed 6/1/2008.

159. See <http://www.scotoma.org/notes/index.cgi?LondonPL>, accessed 6/1/2008.

160. Sigmund Freud, "Fetishism," in *Sexuality and the Psychology of Love*, ed. Philip Rieff (New York: Collier Books, 1963), 205. Freud also writes: "It is not true that the child emerges from his experience of seeing the female parts with an unchanged belief in the woman having a phallus. He retains this belief but he also gives it up; during the conflict between the deadweight of the unwelcome perception and the force of the opposite wish, a compromise is constructed such as is only possible in the realm of unconscious thought—by the primary processes. In the world of psychical reality the woman still has a penis in spite of it all, but this penis is no longer the same as it once was. Something else has taken its place, has been appointed as its successor, so to speak, and now absorbs all the interest which formerly belonged to the penis" (Ibid., 206).

161. William Pietz, "The Problem of the Fetish," pt. 1, *Res: Anthropology and Aesthetics* 9 (Spring 1985): 12.

162. Freud, "Fetishism," 208.

163. Slavoj Žižek, *The Sublime Object of Ideology* (London and New York: Verso, 1989), 31.

164. Alan Turing, "Computing Machinery and Intelligence," *Mind* 59 (1950), <http://www.loebner.net/Prizef/TuringArticle.html>, accessed 3/1/2006.

165. N. Katherine Hayles develops this theme of revealing codes in *My Mother Was a Computer*, 54–61. Importantly, some software art projects also complicate and frustrate code as Xray vision and connection as meaning, such as Golan Levin's *AxisApplet* produced for the Whitney Artport CODeDoc project.

166. Keenan, "The Point Is to (Ex)Change It," 173.

Computers that Roar

1. See John von Neumann, "The First Draft Report on the EDVAC," *Contract No. W–670–ORD–4926* between the U.S. Ordnance (USORD) and the University of Pennsylvania, June 30, 1945, <http://qss.stanford.edu/~godfrey/vonNeumann/vnedvac.pdf>, accessed 9/7/2009; and

Jon Agar, *The Government Machine: A Revolutionary History of the Computer* (Cambridge, Mass.: MIT Press 2003), 391.

2. Paul N. Edwards, *The Closed World: Computers and the Politics of Discourse in Cold War America* (Cambridge, Mass.: MIT Press, 1997); David Golumbia, *The Cultural Logic of Computation* (Cambridge, Mass.: Harvard University Press, 2009); Joseph Weizenbaum, *Computer Power and Human Reason: From Judgment to Calculation* (San Francisco: W. H. Freeman, 1976), 157.

3. Weizenbaum, *Computer Power and Human Reason*, 157.

4. Aristotle as quoted and translated by Paul Ricoeur, *The Rule of Metaphor* (London and New York: Routledge), 13.

5. George Lakoff and Mark Johnson, *Metaphors We Live By* (Chicago: University of Chicago, 1980), 5; emphasis in original.

6. Ibid., 115.

7. They write, "Our ordinary conceptual system, in terms of which we both think and act, is fundamentally metaphorical in nature. The concepts that govern our thought are not just matters of the intellect. They also govern our everyday functioning down to the most mundane details. Our concepts structure what we perceive, how we get around in the world, and how we relate to other people. Our conceptual system thus plays a central role in defining our everyday realities" (Ibid., 3).

8. Ibid., 119.

9. Ibid., 144.

10. Ibid., 239.

11. Ibid., 193.

12. Thomas Keenan, "The Point Is to (Ex)Change It: Reading *Capital* Rhetorically," in *Fetishism as Cultural Discourse*, ed. Emily Apter and William Pietz (Ithaca, N.Y.: Cornell University Press, 1993), 157.

13. Roman Jackobson of course defined *metaphor* in terms of selection and combination—the ability to think through substitutions in "Two Aspects of Language and Two Types of Aphasic Disturbances," *Fundamentals of Language*, 2nd rev. ed. (The Hague: Mouton, 1971).

14. Ricoeur, *Rule of Metaphor*, 23.

15. Ibid., 23–24.

16. Ibid., 44. He writes, "If metaphor belongs to an heuristic of thought, could we not imagine that the process that disturbs and displaces a certain logical order, a certain conceptual hierarchy, a certain classification scheme, is the same as that from which all classification proceeds? . . . could we not imagine that the order itself is born in the same way that it changes? Is there not,

in Gadamer's terms, a 'metaphoric' at work at the origin of logical thought, at the root of all classification? . . . The idea of an initial metaphorical impulse destroys the[se] oppositions between proper and figurative, ordinary and strange, order and transgression. It suggests the idea that order itself proceeds from the metaphorical constitution of semantic fields, which themselves give rise to genus and species" (Ibid., 24).

17. Ibid., 37.

18. Ibid., 48.

2 Daemonic Interfaces, Empowering Obfuscations

1. Luc Boltanski and Eve Chiapello have outlined this encyclopedically in *The New Spirit of Capitalism*, trans. Gregory Elliott (London: Verso, 2005).

2. *Oxford English Dictionary Online*, June 2010, Draft Entry, S.V. "daemon, *n*," emphasis in original.

3. Paul Edwards, *The Closed World: Computers and the Politics of Discourse in Cold War America* (Cambridge, Mass.: MIT Press, 1996), 111, 107.

4. Ibid., 12.

5. Ibid., 13.

6. See David Mindell, *Between Human and Machine: Feedback, Control, and Computing before Cybernetics* (Baltimore, Md.: Johns Hopkins University Press, 2002).

7. See Gordon Bell, "Toward a History of (Personal) Workstations," in *A History of Personal Workstations*, ed. Adele Goldberg (New York: ACM Press; Reading, Mass.: Addison-Wesley Pub. Co., 1988), 29; Martin Campbell-Kelly and William Aspray, *Computer: A History of the Information Machine* (Boulder, Colo.: Westview Press, 2004), 168.

8. As quoted by Edwards, *Closed World*, 258.

9. John von Neumann, *Papers of John von Neumann on Computing and Computer Theory*, ed. William Aspray and Arthur Burks (Cambridge, Mass.: MIT Press, 1987), 413.

10. J. C. R. Licklider, "Man-Computer Symbiosis," in *NewMediaReader*, ed. Noah Wardrip-Fruin and Nick Montfort (Cambridge, Mass.: MIT Press, 2003), 75.

11. As quoted in Joseph Weizenbaum, *Computer Power and Human Reason: From Judgment to Calculation* (San Francisco: W. H. Freeman and Company, 1976), 246.

12. Ben Shneiderman, "Direct Manipulation: A Step Beyond Programming Languages," in *NewMediaReader*, ed. Wardrip-Fruin and Montfort, 486.

13. Michel Foucault, *The Birth of Biopolitics: Lectures at the Collège de France, 1978–1979*, trans. Grahm Burchell (Basingstoke, England, and New York: Palgrave Macmillan, 2008), 252.

14. Boltanski and Chiapello, *New Spirit of Capitalism*, 92.

15. Catherine Malabou, *What Should We Do with Our Brains*, trans. Sebastian Rand (New York: Fordham University Press, 2008), 44.

16. Ibid., 51.

17. George Lakoff and Mark Johnson, *Metaphors We Live By* (Chicago: Chicago University Press, 1980), 70. Ben Shneiderman also directly acknowledges Piaget among others in "Direct Manipulation," 492–493.

18. Lakoff and Johnson, *Metaphors We Live By*, 123.

19. Shneiderman, "Direct Manipulation," 491.

20. Indeed, Laurel argues in *Computers as Theater* (Reading, Mass.: Addison-Wesley Publishers, 1991) that interface metaphors are "dangerous" because they are always doomed to fail: they are similes rather than metaphors, like reality only different (129). This difference can become overwhelming for the user—the interface can become a visible impediment—because the interface refers to the wrong thing: to reality rather than to compelling causality or to action (131).

21. Ibid., xviii.

22. Ibid., 77.

23. Ibid., 33.

24. Ibid., 6; emphasis in original.

25. Edwards draws from Sherman Hawkins's use of the term "to define one of the major dramatic spaces in Shakespearean plays. Closed-world plays are marked by a unity of place, such as a walled city or the interior of a castle or house. Action within this space centers [on] attempts to invade and/or escape its boundaries. Its archetype is the siege , , , the closed world includes not just the sealed, claustrophobic spaces metaphorically marking its closure, but the entire surrounding field in which the drama takes place" (Edwards, *Closed World*, 12–13).

26. She contends, "There are ways in which art is "lawful"; that is, there are formal, structured, and causal dimensions that can be identified and used both descriptively and productively" (Laurel, *Computers as Theater*, 28).

27. Ibid., 67.

28. Ibid., 62.

29. Ibid., 101.

30. Ibid., 167.

31. See Wendy Hui Kyong Chun, *Control and Freedom: Power and Paranoia in the Age of Fiber Optics* (Cambridge, Mass.: MIT Press, 2006).

32. Louis Althusser, "Ideology and Ideological State Apparatuses (Notes Towards an Investigation)," in *Lenin and Philosophy and Other Essays*, trans. Ben Brewster (New York: Monthly Review Press, 2001), 109.

33. Ibid., 171 (emphasis in original).

34. See Slavoj Žižek, *The Sublime Object of Ideology* (London and New York: Verso, 1989), 11–53.

35. See Ceruzzi, *A History of Modern Computing*, 208–209; Martin Campbell-Kelly, *From Airline Reservations to Sonic the Hedgehog* (Cambridge, Mass.: MIT Press), 207–229.

36. See Thomas Y. Levin, "Rhetoric of the Temporal Index: Surveillant Narration and the Cinema of 'Real Time,'" in *CTRL Space: Rhetorics of Surveillance from Bentham to Big Brother*, ed. Thomas Y. Levin et al. (Cambridge, Mass.: MIT Press, 2002), 578–593.

37. Tara McPherson, "Reload: Liveness, Mobility and the Web," in *The Visual Culture Reader*, 2nd ed., ed. Nicholas Mirzoeff (London and New York: Routledge, 2002), 461–462.

38. Coming from film rather than from television studies and focusing more on applications than on phenomenology, Alexander Galloway in *Protocol: How Power Exists after Decentralization* (Cambridge, Mass.: MIT Press, 2004) similarly argues that continuity makes websurfing "a compelling, intuitive experience for the user":

On the Web, the browser's movement is experienced as the user's movement. The mouse movement is substituted for the user's movement. The user looks through the screen into an imaginary world, and it makes sense. The act of "surfing the web," which, phenomenologically, should be an unnerving experience of radical dislocation—passing from a server in one city to a server in another city—could not be more pleasurable for the user. Legions of computer users live and play online with no sense of radical dislocation. (64)

39. Lev Manovich, "Generation Flash," 2002, <http://www.manovich.net/DOCS/generation_flash.doc>, accessed 8/8/2010.

40. Julian Dibbell, *My Tiny Life: Crime and Passion in a Virtual World* (New York: Holt, 1998).

41. Manovich, "Generation Flash," 2002.

42. Fredric Jameson, *Postmodernism, or the Cultural Logic of Late Capitalism* (Durham, N.C.: Duke University Press, 1991), 51.

43. However, these parallels arguably reveal the fact that our understandings of ideology are lacking precisely to the extent that they, like interfaces, rely on a fundamentally theatrical model of behavior.

44. This argument thus seeks to complicate Matthew Kirschenbaum's insightful critique in *Mechanisms: New Media and the Forensic Imagination* (Cambridge, Mass.: MIT Press, 2008) of "medial ideology"—the acceptance of the interface as the computer—rampant in new media studies (36–45). This medial ideology is so attractive not simply because we are enamored by the flickering signifiers on our screens, but also because interfaces offer us a way to "map" our larger relation to the world, to ideology itself.

45. See Jean François Lyotard, *The Postmodern Condition: A Report on Knowledge* (Minneapolis: University of Minnesota Press, 1984).

46. Jameson, *Postmodernism*, ix.

47. Ibid., x. Formally, Jameson argues, postmodern art—which erodes the barrier between high and low culture—has five characteristics: (1) a new depthlessness, in which inimitable styles are turned into surface characteristics (postmodern codes, blank parodies or "pastiche"); (2) the waning of affect (the waning of the subject and of the difference between inside and outside); (3) a weakening of historicity (of the relationship between past, present, and future); (4) a change in tone (toward the sublime); and (5) "a constitutive relationship of all this to a whole new technology, which is itself a reflection, or a way to deal with a whole new economic world" (*Postmodernism*, 6).

Following from Jameson's description of postmodernism—in particular from his likening of postmodernism to pastiche and his view of technology as reflection—Lev Manovich, in *The Language of New Media* (Cambridge, Mass.: MIT Press, 2001), has argued that GUIs are quintessentially postmodern. It is no accident, he writes, that the GUI, "which legitimized a 'cut and paste' logic, as well as media manipulation software such as Photoshop, which popularized a plug-in architecture, took place during the 1980s—the same decade when contemporary culture became 'postmodern'" (131). The GUI's cut-and-paste logic, Manovich argues, is emblematic of a culture that

no longer tried to "make it new." Rather, endless recycling and quoting of past media content, artistic styles, and forms became the new "international style" and the new cultural logic of modern society. Rather than assembling more media recordings of reality, culture is now busy reworking, recombining, and analyzing already accumulated media material. Invoking the metaphor of Plato's cave, Jameson writes that postmodern cultural production "can no longer look directly out of its eyes at the real world but must, as in Plato's cave, trace its mental images of the world on its confining walls." In my view, this new cultural condition found its perfect reflection in the emerging computer software of the 1980s that privileged selection from ready-made elements over creating them from scratch. And to a large extent it is this software that in fact made postmodernism possible. (Ibid.)

Manovich's application of Jameson's diagnosis of postmodernism to computer interfaces is intriguing, although, given Lyotard's linking of postmodernism to new creative acts, arguably one-sided and reductive. Lyotard links postmodernism to paralogy and the sublime—a challenge to totalitarianism. More important, however, by focusing on formal likenesses, Manovich misses the larger point: rather than a symptom or a cause, GUIs, which emerged as a common cultural object much later than Jameson's and Lyotard's initial descriptions of postmodernism in the 1970s, are a *response* to the challenges of postmodernism, to the spatial challenges it posed. Interfaces, that is, are ways to navigate postmodern confusion.

48. Fredric Jameson, "Cognitive Mapping," in *Marxism and the Interpretation of Culture*, ed. Cary Nelson and Lawrence Grossberg (Champaign: University of Illinois Press, 1988), 351.

49. Ibid., 351.

50. Jameson, *Postmodernism*, 39.

51. It stems from an earlier disorientation, brought about by global imperialism, in which the truth of a subject's experience

no longer coincides with the place in which it takes place. The truth of that limited daily experience of London lies, rather, in India or Jamaica or Hong Kong; it is bound up with the whole colonial system of the British Empire. . . . Yet those structural coordinates are no longer accessible to immediate lived experience and are often not even conceptualizable for most people.

   There comes into being, then, a situation in which we can say that if individual experience is authentic, then it cannot be true, and that if a scientific or cognitive model of the same content is true, then it escapes individual experience. (Jameson, "Cognitive Mapping," 349)

52. Ibid., 353.

53. Jameson argues, "the conception of capital is admittedly a totalizing or systemic concept: no one has ever seen or met the thing itself; it is either the result of scientific reduction (and it should be obvious that scientific thinking always reduces the multiplicity of the real to a small-scale model) or the mark of an imaginary and ideological vision" ("Cognitive Mapping," 354).

54. Ibid., 356.

55. Jameson, *Postmodernism*, 37.

56. Ibid., 54.

57. For an excellent analysis of free labor, see Tiziana Terranova's *Network Culture: Politics for the Information Age* (London: Pluto Press, 2004).

58. David Harvey, *A Brief History of Neoliberalism* (Oxford: Oxford University Press, 2005), 3.

59. See Foucault, *Birth of Biopolitics*, 280.

60. David Mindell similarly argues, "Our computers retain traces of earlier technologies, from telephones and mechanical analogs to directorscopes and tracking radars. . . . When we articulate a mouse to direct a machine, do we not resemble Sperry's pointer-matching human servomechanisms? When we interpret glowing images and filter out signals from noise, do we not resemble a pip-matching radar operator?" See Mindell, *Between Human and Machine*, 321.

61. See Linda C. Smith, "Memex as an Image of Potentiality Revisited," in *From Memex to Hypertext: Vannevar Bush and the Mind's Machine*, ed. James M. Nyce and Paul Kahn (Boston: Academic Press, 1991), 261–286.

62. Indeed, Vannevar Bush deliberately contrasts the memex to expensive digital computers in "Memex Revisited," in *New Media, Old Media*, ed. Wendy Hui Kyong Chun and Thomas Keenan (New York: Routledge, 2006), 86–96.

63. Vannevar Bush, "As We May Think," *The Atlantic* (July 1945), <http://www.theatlantic.com/doc/194507/bush>, accessed 9/9/2009.

64. Vannevar Bush, "Memorandum Regarding Memex," *From Memex to Hypertext*, 81.

65. Bush, "As We May Think," n.p.

66. For Nelson and Engelbart, the "gadgetry" Bush envisions guarantees freedom. Nelson, in his "As We Will Think" (in *From Memex to Hypertext: Vannevar Bush and the Mind's Machine*, ed. James M. Nyce and Paul Kahn [Boston: Academic Press, 1991]) pinpoints Bush's notion of a "trail" as hypertext, where hypertext more generally means a "text structure that cannot be conveniently printed" (Ibid., 253). Hypertext, Nelson stresses, was to liberate human thinking: "Let me suggest that such an object and system [hypertext], properly designed and administered, could have great potential for education, increasing the student's range of choices, his sense of freedom, his motivation, and his intellectual grasp." See Theodor Nelson, "A File Structure for the Complex, the Changing, and the Indeterminate," in *NewMediaReader*, ed. Wardrip-Fruin and Montfort, 144). Engelbart, focusing on the manipulation of symbols, argues, in relation to his own system, "you are quite elated by this freedom to juggle your thoughts, and by the way this freedom allows you to *work* them into shape." See Douglas C. Engelbart, "Augmenting Human Intellect: A Conceptual Framework" (October 1962), <http://www.dougengelbart.org/pubs/augment-3906.html>, accessed 8/8/2009; emphasis in original. Bush himself contends that *the* problem hindering scientists and scientific progress is access: man must "mechanize his records more fully if he is to push his experiment [human civilization] to its logical conclusion and not merely become bogged down part way there by overtaxing his limited memory" ("As We May Think," section 8, n.p.).

67. Bush, "As We May Think," section 1, n.p.

68. Ibid.

69. Bush, "Memex Revisited," 85.

70. Ibid., 91, 93, 94.

71. Bush, "As We May Think," section 6, n.p.

72. Ibid., sections 6, 8, n.p.

73. Ibid., section 4, n.p.; Bush, "Memex Revisited," 85.

74. This is Foucault's diagnosis of traditional history in *The Archaeology of Knowledge & The Discourse on Language*, trans. A. M. Sheridan Smith (New York: Pantheon Books, 1982).

75. Noah Wardrip-Fruin, "Introduction to 'As We May Think,'" in *NewMediaReader*, ed. Wardrip-Fruin and Montfort, 35; emphasis in original.

76. Jann Sapp, "The Nine Lives of Gregor Mendel," *Experimental Inquiries: Historical, Philosophical, and Social Studies of Experimentation in Science*, ed. H. E. Le Grand (Dordrecht, The Netherlands: Kluwer Academic Publishers, 1990), 137–166.

77. See Jacques Derrida, *Archive Fever: A Freudian Impression*, trans. Eric Prenowitz (Chicago: University of Chicago Press, 1996). The pleasure of forgetfulness is to some extent the pleasure of death and destruction. It is thus no accident that this "supplementing" of human memory has also been imagined as the death of the human species in so many fictions and films, and that déjà vu is the mark of the artificial in the Wachowski brothers' film *The Matrix*.

78. Douglas C. Engelbart, "Letter to Vannevar Bush and Program On Human Effectiveness," *From Memex to Hypertext*, 236.

79. Douglas C. Engelbart, "The Augmented Knowledge Workshop," in *Proceedings of the ACM Conference on the History of Personal Workstations* (New York: ACM Press, 1986), 74.

80. Engelbart, "II: Conceptual Framework," in "Augmenting Human Intellect," n. p.

81. Engelbart, "1.A, General Introduction," in "Augmenting Human Intellect," n. p.

82. Engelbart, "II: Conceptual Framework," n. p.

83. As quoted by Thierry Bardini, in *Bootstrapping: Douglas Engelbart, Coevolution, and the Origins of Personal Computing* (Stanford, Calif.: Stanford University Press, 2000), 18.

84. Ibid., 19.

85. Engelbart, "Hypothetical Description of Computer-Based Augmentation System," in "Augmenting Human Intellect," n. p.

86. For more on this see Foucault, *Birth of Biopolitics*, 243–246. See also Philip Agre's discussion of capture as making more tasks intelligible as market transactions in "Surveillance and Capture: Two Models of Privacy," *The Information Society* 10, no. 2 (1994): 101–127.

87. For more on this, see Jane Feuer, "The Concept of Live Television: Ontology as Ideology," in *Regarding Television: Critical Approaches*, ed. E. Ann Kaplan (Washington, D.C.: University Press of America, 1983), 12–22.

88. Cornelia Vismann, *Files: Law and Media Technology*, trans. Geoffrey Winthrop-Young (Stanford, Calif.: Stanford University Press, 2008), 6.

89. Ibid., 7.

90. Ibid.,10.

91. Ibid., 163.

92. The following PERL program, for instance, says hello every 5 minutes:

```
use POSIX qw(setsid);

#turns the process into a session leader, group leader, and ensures that it doesn't #have a
controlling terminal

chdir '/'            or die "Can't chdir to /: $!";
umask 0;
open STDIN, '/dev/null' or die "Can't read /dev/null: $!";
#open STDOUT, '>/dev/null' or die "Can't write to /dev/null: $!";
open STDERR, '>/dev/null' or die "Can't write to /dev/null: $!";
defined(my $pid = fork) or die "Can't fork: $!";
```

```
exit if $pid;
setsid              or die "Can't start a new session: $!";
while(1) {
sleep(5);
print "Hello...\n";
}
```

From <http://www.webreference.com/perl/tutorial/9/3.html>, accessed 9/1/2008.

93. Fernando J. Corbato, as quoted in "The Origin of the Word Daemon," from Richard Steinberg/Mr. Smarty Pants, *The Austin Chronicle*, <http://ei.cs.vt.edu/~history/Daemon.html>, accessed 3/1/2007. This is why Neal Stephenson in *Snow Crash* (New York: Bantam Books, 1993) describes robots or servants in the Metaverse as daemons.

94. This resonates with Derrida's analysis of writing, in contrast to "living logos," as "orphaned" in "Plato's Pharmacy," *Dissemination*, trans. Barbara Johnson (Chicago: University of Chicago, 1981), 76.

95. John Schwartz, "Privacy Fears Erode Support for a Network to Fight Crime," *New York Times*, March 15, 2004, <http://www.nytimes.com/2004/03/15/technology/15matrix.html?pagewanted=1>, accessed 4/15/2004.

96. Ibid.

97. Stephanie Clifford, "Cable Companies Target Commercials to Audience," *New York Times*, March 3, 2009, <http://www.nytimes.com/2009/03/04/business/04cable.html>, accessed 9/9/2009.

98. "'Personal data' in iTunes tracks," *BBC News Online*, June 1, 2007, <http://news.bbc.co.uk/2/hi/technology/6711215.stm>, accessed 8/1/2008.

99. Manovich, *Language of New Media*, 48.

100. Microsoft is considering such actions in its Palladium initiative.

101. See Friedrich Kittler, "There Is No Software," *Ctheory.net*, October 18, 1995, <http://www.ctheory.net/articles.aspx?id=74>, accessed 8/8/2010.

102. See Francois Jacob, *The Logic of Life: A History of Heredity*, trans. Betty E. Spillman (New York: Pantheon Books, 1973), 247–298.

103. Richard Doyle's concept of "rhetorical software," developed in his *On Beyond Living: Rhetorical Transformations of the Life Sciences* (Stanford, Calif.: Stanford University Press, 1997), is emblematic of the use of *software* as a critical term in nonscientific scholarly discourse.

104. Bruce Schneier, "U.S. Enables Chinese Hacking of Google," *CNN Opinion Online*, <http://www.cnn.com/2010/OPINION/01/23/schneier.google.hacking/index.htm>, accessed 3/28/2010.

105. Ibid.

106. "Government Information Awareness," *SourceWatch.org*, <http://www.sourcewatch.org/index.php?title=Government_Information_Awareness>, accessed 3/28/2010.

107. Google, "Explore Flu Trends Around the World," <http://www.google.org/flutrends/>, accessed June 24, 2009.

108. Adrian Mackenzie, *Cutting Code: Software and Sociality* (New York: Peter Lang Pub. Inc., 2006), 169.

109. Milton Friedman, *Capitalism and Freedom*, Fortieth Anniversary Edition (Chicago: Chicago University Press, 2002), 30.

110. Friedrich Nietzsche, "On Truth and Lie in an Extra-Moral Sense" (1873), from the *Nachlass*, trans. Walter Kaufmann and Daniel Breazeale <www.geocities.com/thenietzschechannel/tls.htm>, accessed 9/3/2009.

II Regenerating Archives

1. Wolfgang Ernst, "Art of the Archive," *Künstler.Archiv—Neue Werke zu historischen Beständen*, ed. Helen Adkins (Köln: Walter König, 2005), 99.

2. Jacques Derrida, "Freud and the Scene of Writing," *Writing and Difference*, trans. Alan Bass (Chicago: University of Chicago, 1978), 226.

3. Execution, as this book has been arguing, is arguably a more critical category, but it is constantly overlooked in favor of memory.

4. The move from calculator to computer is also the move from mere machine to human-emulator: IBM initially resisted the term *computer* because computers initially were human. To call a machine a computer implied job redundancy (Martin Campbell-Kelly and William Aspray, *Computer: A History of the Information Machine* (New York: Basic Books, 1996), 115). John von Neumann, in his mythic and controversial *The First Draft Report of the EDVAC* (1945) deliberately used the term *memory organ* rather than *store*, also in use at the time, in order to parallel biological and computing components and to emphasize the ephemeral nature of vacuum tubes. (See von Neumann, "First Draft of a Report on the EDVAC," <www.cs.colorado.edu/~zathras/csci3155/EDVAC_vonNeumann.pdf>, accessed 9/12/2003). Vacuum tubes, unlike mechanical switches, can hold values precisely because their signals can degenerate—and thus regenerate.

5. For more on this compression see David Harvey, *A Short History of Neoliberalism* (Oxford: Oxford University Press, 2005, 7).

6. See Ryan Lizza, "The YouTube Election," *New York Times*, August 20, 2006, <http://www.nytimes.com/2006/08/20/weekinreview/20lizza.html?ex=1313726400&en=a605fabfcb81eebf&ei=5088&partner=rssnyt&emc=rss>, accessed 9/1/2009. In this video clip Senator George Allen referred to an Indian American man as a "macacca" while campaigning in Virginia.

7. For more on this ideal and its incapacity to explain public behavior, see Thomas Keenan's "Publicity and Indifference (Sarajevo on Television)," *PLMA* 111, no. 1 (January 2002): 104–116.

8. Jacques Derrida, *Archive Fever: A Freudian Impression*, trans. Eric Prenowitz (Chicago: Chicago University Press, 1998), 11–12.

9. Derrida, *Archive Fever*, 84.

10. Derrida, "Freud and the Scene of Writing," 228.

11. Howard Caygill, "Meno and the Internet: Between Memory and the Archive," *History of the Human Sciences* 12 (1999): 2.

12. Derrida, *Archive Fever*, 36.

13. Ibid.

14. Ibid., 4n1.

15. For the relationship between threat and promise see Jacques Derrida, "Typewriter Ribbon," *Without Alibi* (Palo Alto, Calif.: Stanford University Press, 2002), 155.

16. Ibid., 7.

17. Ibid., 1; emphasis in original.

18. According to Ernst, archives are comprised of fundamentally discontinuous units or fragments—they are not histories (stories), but rather "discrete, isolated units and islands of discourse" that are "generally smoothed over in narrative representations based on archive research" (Ernst, "Art of the Archive," 94). These discrete units do not simply lie about, but rather are ordered—the archive "is a metonymic device that 'formalizes' experience"; "the art of the archive entails assigning names to documents—to which stories immediately adhere" (Ibid., 94, 96). In contrast to this metonymic linguistic ordering based on names and categories, digital media offers the possibility of searching based on textual and pattern-based similarities, disrupting archival ordering. This disruption undermines a fundamental function of the archive: the separation of documents into discrete units; see Wolfgang Ernst, "Dis/continuities: Does the Archive Become Metaphorical in Multi-Media Space?," in *New Media, Old Media: A History and Theory Reader*, ed. Wendy Hui Kyong Chun and Thomas Keenan (New York: Routledge, 2006), 119. Rather than an archive, one has something like a "life stream"—a time based medium that is only superficially comprised of documents. In addition, cyberspace has no memory: "the archival phantasms in cyberspace are an ideological deflection of the sudden erasure of archives (both hard- and software) in the digital world. . . . The Internet has no organized memory and no central agency, being defined rather by the circulation of discrete states. If there is memory, it operates as a radical constructivism: always just situationally built, with no enduring storage" (Ibid., 119). This constant flow and information in general, whose value is linked to entropy rather than order technically redeems (wipes out) the archive (Ibid., 97). "The archive is a given, well-defined-lot; the Internet, on the contrary, is not just a collection of unforeseen texts, but of sound and images as well, an archive of sensory data for which no genuine archival culture has been developed so far

in the occident. . . . What separates the Internet from the classical archive is that its mnemonic logic is more dynamic than cultural memory in the printed archive" (Ibid., 119–120).

3 Order from Order, or Life According to Software

1. Pierre-Simon Laplace, *A Philosophical Essay on Probabilities* (1820; reprinted, New York: Dover, 1951), preface.

2. Gilles Deleuze and Félix Guattari, *A Thousand Plateaus: Capitalism and Schizophrenia*, trans. Brian Massumi (Minneapolis: University of Minnesota Press, 1987), 143.

3. See Paul Edwards, *The Closed World: Computers and the Politics of Discourse in Cold War America* (Cambridge, Mass.: MIT Press, 1996); and Richard Boyd, "Metaphor and Theory Change: What Is a 'Metaphor' a Metaphor For?," in *Metaphor and Thought*, ed. Andrew Ortony (Cambridge: Cambridge University Press, 1993), chap. 21.

4. Richard Dawkins, *The Selfish Gene*, 2nd ed. (Oxford: Oxford University Press, 1989), v.

5. Ibid., 49.

6. François Jacob, *The Logic of Life* (New York: Pantheon, 1982), 2.

7. John von Neumann, "First Draft of a Report on the EDVAC," Contract No. W-670-ORD-4926 between the United States Army Ordnance Department and the University of Pennsylvania, June 30, 1945, <http://qss.stanford.edu/~godfrey/vonNeumann/vnedvac.pdf>, 3, accessed 8/8/2010.

8. Herman Goldstine and John von Neumann, "Planning and Coding of Problems for an Electronic Computer Instrument," "Report on the Mathematical and Logical Aspects of an Electronic Computing Instrument," Part II, Volume I (Princeton, N.J.: Institute for Advanced Study, 1947).

9. The other factors, such as clerical and gendered bureaucratic relations, and Fordist industrial techniques, are the subjects of the other chapters.

10. Jacob, *Logic of Life*, 264–265.

11. Ibid., 1.

12. Lily Kay, *Who Wrote the Book of Life? A History of the Genetic Code* (Stanford, Calif.: Stanford University Press, 2000), xv.

13. Ibid., 85.

14. Richard Doyle, *On Beyond Living: Rhetorical Transformations of the Life Sciences* (Stanford, Calif.: Stanford University Press, 1997), 1.

15. Ibid., 13.

16. Ibid., 3.

17. Ibid., 2.

18. Ibid., 6–7.

19. David Mindell in *Between Human and Machine: Feedback, Control, and Computing Before Cybernetics* (Baltimore: Johns Hopkins University Press, 2002) argues convincingly against this claim, emphasizing the importance of electrical engineering to the development of cybernetics.

20. Norbert Wiener, *The Human Use of Human Beings: Cybernetics and Society* (Garden City, N.Y.: Doubleday, 1954), 27; emphasis in original.

21. Wiener explains, "Information is a name for the content of what is exchanged with the outer world as we adjust to it, and make our adjustment felt upon it. The process of receiving and of using information is the process of our adjusting to the contingencies of the outer environment, and of our living effectively within that environment" (*Human Use of Human Beings*, 17–18).

22. On the importance of the military, see Heinz von Foerster, "Epistemology of Communication," in *The Myths of Information: Technology and Postindustrial Culture*, ed. Kathleen Woodward (Madison, Wisc.: Coda Press, 1980), 18–27.

23. Norbert Wiener, *Cybernetics, or, Control and Communication in the Animal and the Machine* (Cambridge, Mass.: Technology Press, 1948), 118. Paul Edwards in *The Closed World* similarly distinguishes between cybernetic and artificial intelligence in terms of hardware versus software as the means for modeling brains (239).

24. Warren Weaver, "Some Recent Contributions," in Claude E. Shannon and Warren Weaver, *The Mathematical Theory of Communication* (Urbana: University of Illinois Press, 1963), 18.

25. Claude Shannon, "The Mathematical Theory of Communication," in Shannon and Weaver, *Mathematical Theory of Communication*, 31; emphasis in original.

26. Jacob, *Logic of Life*, 79, 207.

27. Ibid., 298.

28. Richard Panek, *The Invisible Century: Einstein, Freud, and the Search for Hidden Universes* (New York: Penguin, 2004); emphasis in original. Panek theorizes the invisible century in terms of the Freudian unconscious and Einsteinian relativity. In contrast, genetics and computer memory do not simply speculate about the invisible, they posit an invisible program that drives the visible world.

29. Erwin Schrödinger, *What Is Life?* with *Mind and Matter* and *Autobiographical Sketches* (Cambridge, UK: Cambridge University Press, 1992), 23; and Jacob, *Logic of Life*, 285.

30. Schrödinger, *What Is Life?*, 68–69.

31. Ibid., 4.

32. Ibid., 31.

33. Ibid., 69.

34. Ibid., 73–74.

35. Schrödinger himself rejected such a connection. See Kay, *Who Wrote the Book of Life?*, 64–65.

36. Kay, *Who Wrote the Book of Life?*, 21.

37. Schrödinger, *What Is Life?*, 23.

38. Schrödinger to E. I. Conway, October 25, 1942, quoted in E. J. Yoxen, "The Social Impact of Molecular Biology" (unpublished PhD thesis, Cambridge University, 1978), 152.

39. Schrödinger, *What Is Life?*, 23.

40. Ibid., 21–22. By making the code-script both pattern and development, Schrödinger possibly was rejecting the classical genetic separation of transmission from development, but since his code-script conflates development with the means of transmission, it hardly offers a serious biological consideration of development.

41. Kay, *Who Wrote the Book of Life?*, xvii.

42. Kay similarly argues that a poststructuralist view of writing sees writing as writing itself. My argument is slightly different: it's not the poststructuralist view but rather the view that writing and execution are conflated, although importantly, this formulation also gives great powers to anyone who controls the source.

43. Jacob, *Logic of Life*, 298.

44. François Jacob, *The Statue Within: An Autobiography* (New York: HarperCollins, 1988), 58.

45. Linus Pauling, "Schrödinger's Contributions to Chemistry and Biology," in *Schrödinger: Centenary Celebrations of a Polymath*, ed. C. W. Kilmister (Cambridge: Cambridge University Press, 1987), 225–233.

46. Kay argues, "Schrödinger's code-script was based on permutations in proteins, it neither related one system of symbols . . . to another . . . as did genetic codes after 1953, nor, most importantly did it claim to transfer information" (*Who Wrote the Book of Life?*, 61.)

47. Ibid., 161.

48. Gunther S. Stent, "Introduction: Waiting for the Paradox," in *Phage and the Origins of Molecular Biology*, ed. John Cairns et al. (Long Island, N.Y.: Cold Spring Harbor Laboratory of Quantitative Biology, 1966), 3.

49. Donald Fleming, "Émigré Physicists and the Biological Revolution," in *The Intellectual Migration*, ed. Donald Fleming and Bernard Bailyn (Cambridge, Mass.: Harvard University Press, 1969), 172.

50. Evelyn Fox Keller, "Physics and the Emergence of Molecular Biology: A History of Cognitive and Political Synergy," *Journal of the History of Biology* 23, no. 3 (Fall 1990): 390.

51. Leah Ceccarelli, *Shaping Science with Rhetoric: The Cases of Dobzhansky, Schrödinger, and Wilson* (Chicago: University of Chicago, 2001), 82–112.

52. Ibid., 67.

53. Doyle, *On Beyond Living*, 13.

54. Ibid., 28–29.

55. Ibid., 13; emphasis in original.

56. Ibid., 35. Not surprisingly, this part of Schrödinger's text has been mainly forgotten, since his hypothesis has proven so incorrect. What Schrödinger could not foresee is not only the importance of hydrogen bonds, but also the complete conflation of message with action, necessary for code-script as life.

57. Michel Foucault, *The Archaeology of Knowledge & The Discourse on Language*, trans. A. M. Sheridan Smith (New York: Pantheon Books, 1972), 129.

58. Ibid.; emphasis in original.

59. Michel Foucault, *The Order of Things: An Archaeology of the Human Sciences*, trans. A. M. Sheridan (New York: Vintage Books, 1994), ix.

60. Ibid., 131.

61. Ibid., 15.

62. Deleuze argues in *Foucault*, trans. Sean Hand (Minneapolis: University of Minnesota, 1988), "each age has its own particular way of putting language together, because of its different groupings" (56) and "each historical formation sees and reveals all it can within the conditions laid down for visibility, just as it says all it can within the conditions relating to statements" (59).

63. Foucault, *Archaeology of Knowledge*, 128, and *Order of Things*, xx.

64. Foucault, *Order of Things*, ix.

65. See Michel Foucault, "Two Lectures," in *Power/Knowledge: Selected Interviews and Other Writings, 1972–1977*, ed. Colin Gordon (New York: Pantheon, 1980), 78–108, and "What Is Critique," in *What Is Enlightenment? Eighteenth-Century Answers and Twentieth-Century Questions*, ed. James Schmidt (Berkeley, Calif.: University of California Press, 1996), 394–395.

66. Foucault, *Archaeology of Knowledge*, 106.

67. Ibid., 130.

68. See Walter Gilbert, "The RNA World," *Nature* 319 (February 1986): 618.

69. A. M. Turing, "Computing Machinery and Intelligence," *Mind* 59 (1950): 433–460.

70. Ibid., 440.

71. Kay, *Who Wrote the Book of Life?*, 106.

72. John von Neumann, "General and Logical Theory of Automata," in *Papers of John von Neumann on Computing and Computing Theory*, ed. William Aspray and Arthur Burks (Cambridge, Mass.: MIT Press, 1986), 42.

73. H. J. Muller, "Mutation," in *Eugenics, Genetics and the Family, Volume 1. Scientific Papers of the Second International Congress of Eugenics* (Baltimore, Md.: Williams & Wilkins Co., 1923), 106.

74. R. C. Punnett, *Mendelism* (London: Macmillan & Co., 1919).

75. Jan Sapp, "The Nine Lives of Gregor Mendel," in *Experimental Inquiries: Historical, Philosophical and Social Studies of Experimentation in Science*, ed. H. E. Le Grand (Dordrecht/Boston/London: Kluwer Academic Publishers, 1990), 138. The author/date citations included are as they appear in the chapter by Sapp.

76. Ibid., 141.

77. Ibid., 149.

78. Daniel J. Kevles, *In the Name of Eugenics: Genetics and the Uses of Human Heredity* (Cambridge, Mass.: Harvard University Press, 1998), 42.

79. Raphael Falk, "The Real Objective of Mendel's Paper: A Response to Monaghan and Corcos," *Biology and Philosophy* 6, no. 4 (1991): 448.

80. Punnett, *Mendelism*, 32.

81. Francis Galton, "Eugenics: Its Definition, Scope, and Aims," *The American Journal of Sociology* 10, no. 1 (July 1904): 1–6.

82. The British biometricians were more focused on the degeneration of the British race. It is telling that when Fisher changed the subtitle of the journal from "devoted to the genetic study of racial problems" to "devoted to the study of human populations," racial problems referred to problems in one's own race.

83. Kevles, *In the Name of Eugenics*, 9.

84. As quoted in Kevles, *In the Name of Eugenics*, 17.

85. Ronald Fischer, a population geneticist and eugenics advocate, is mainly heralded as breaching the differences between the biometricians and Mendelians. His "evolutionary synthesis" entailed proving that discontinuous changes were not as large as the early Mendelians believed (they were not "sports," which led to new species) and by showing that the continuous traits the biometricians tracked were actually comprised of multiple discrete ones.

86. See Raphael Falk, "What Is a Gene?" *Studies in the History and Philosophy of Science* 17, no. 2 (1986): 138.

87. The drive toward establishing genes as actual physical entities is attributed to Thomas Morgan's lab, in particular to the work of Hermann J. Muller. Morgan's lab, working with drosophila, showed that certain traits were sex-linked, that is, lay on the X or Y chromosome. For Muller—Falk argues in "What Is a Gene?"—genes were units in their own right with inherent characteristics, even if they could only be recognized by their effects. They were a "substance causing reproduction of its own specific composition, but can nevertheless mutate, and retain the property of reproducing itself in various new forms" (Falk, "What Is a Gene," 150). This

material view of the gene loosened the one-to-one correspondence between character and genetic factor, enabling a more modern and logical relationship between the gene and the trait to emerge. It was not until 1944, however, that nucleic acids rather than proteins were revealed as the material core of heredity, and not until the 1950s that there was a consensus on that subject.

88. See Raphael Falk, "The Struggle of Genetics for Independence," *Journal of the History of Biology* 28, no. 2 (1995): 239.

89. See Jan Sapp, "The Struggle for Authority in the Field of Heredity, 1900–1932: New Perspectives on the Rise of Genetics," *Journal of the History of Biology* 16, no. 3 (1983): 322.

90. George Stocking Jr., "The Turn-of-the-Century Concept of Race," *Modernism/Modernity* 1, no. 1 (January 1994): 6.

91. Ibid., 10. Intriguingly, though, this confusion of race and nation had its limitations. Nicholas Hudson argues in "From 'Nation' to 'Race': The Origin of Racial Classification in Eighteenth Century Thought," *Eighteenth-Century Studies* 29, no. 3 (1996) that although race and nation both stem from the same concept of lineage or stock (249), the Africans represented a special case, since they "roughly constituted a single 'race' even in the traditional sense of lineage" (249). As well, in the eighteenth century, the detail devoted to the Native Americans was disparaged as writers emphasized the similarities of the American race, and denied "savages" the complexity of nationality (256). Race as a scientific category more familiar to those in the twentieth century— like disciplinary techniques such as fingerprinting—was first used to describe others, and then applied to "civilized" folk.

92. William Provine, "Geneticists and Race," *American Zoologist* 26 (1986): 859–860.

93. Stocking, "Turn-of-the-Century Concept," 16.

94. Ernst Mayr, *The Growth of Biological Thought: Diversity, Evolution, and Inheritance* (Cambridge, Mass.: Belknap Press, 1982), 695–698.

95. Ibid., 700.

96. Diane Paul, "'In the Interests of Civilization': Marxist Views of Race and Culture in the Nineteenth Century," *Journal of the History of Ideas* 42, no. 1 (January–March 1981): 116–117.

97. See Evelyn Fox Keller, "Nature, Nurture, and the Human Genome Project," in *Code of Codes*, ed. D. J. Kevles and L. Hood (Cambridge, Mass.: Harvard University Press, 1991), 281–357; and Eve Kosofsky Sedgwick, *Tendencies* (New York: Routledge, 1994), 160.

98. Punnett, *Mendelism*, 167–168.

99. Ibid., 169.

100. Ibid.

101. Nils Roll-Hansen, "The Progress of Eugenics: Growth of Knowledge and Change in Ideology," *History of Science* 26 (1988): 293–331.

102. Curt Stern as quoted by Diane Paul, "From Eugenics to Medical Genetics," *Journal of Policy History* 9, no. 1 (1997): 101.

103. Garland Allen, "The Social and Economic Origins of Genetic Determinism: A Case Study of the American Eugenics Movement 1900–1940 and Its Lessons for Today," *Genetica* 99, no. 2–3 (March 1997): 77–88.

104. Michel Foucault, *History of Sexuality*, volume 1, trans. Robert Hurley (New York: Vintage, 1978), 136.

105. Ibid., 137.

106. Michel Foucault, *Security, Territory, Population: Lectures at the College de France 1977–1978*, trans. Graham Burchell (New York: Picador, 2007), 2.

107. Ibid., 79.

108. Charles B. Davenport, *Heredity in Relation to Eugenics* (New York: Arno Press & New York Times, 1972), 265–266.

109. Ibid., 266.

110. For more on genetics as human capital see Michel Foucault, *The Birth of Biopolitics: Lectures at the Collège de France, 1978–1979*, trans. Graham Burchell (Basingstoke, England, and New York: Palgrave Macmillan, 2008), 227.

111. Margaret Sanger, "Chapter XVIII: The Goal," in *Woman and the New Race* (New York: Truth Pub. Co., 1920), 229.

112. Foucault, *History of Sexuality*, 147.

113. Foucault, *Security, Territory, Population*, 100.

114. Michel Foucault, *Society Must Be Defended: Lectures at the College de France 1975–1976*, trans. David Macey (New York: Picador, 2003), 255.

115. Sapp, "Struggle for Authority," 337, 318.

116. Jacob, *Logic of Life*, 1.

117. Ibid., 338.

118. Davenport, *Heredity in Relation to Eugenics*, 1, 2, 4.

119. Ibid., 7; emphasis in original.

120. Ibid., 234; emphasis in original.

121. Ibid., 255.

122. Ibid., 260.

123. See Diane B. Paul and Hamish G. Spencer, "Did Eugenics Rest on an Elementary Mistake?," in *Thinking about Evolution: Historical, Philosophical, and Political Perspectives*, ed. Rama S. Singh et al. (Cambridge: Cambridge University Press, 2000), 103–118.

124. Provine, "Geneticists and Race," 857. Nils Roll-Hansen, responding to Provine's and Paul's arguments, takes a far more positivist position, arguing that the trend, from the new genetics of

1915 on, has been, and continues to be that the more that is known of human heredity, the more ideas of human programming are abandoned. For him, politics did have some role to play in this trajectory, but eugenics, he insists, was abandoned for scientific reasons. Roll-Hansen's argument, however, relies on a rather specious distinction between science and "mere politics" (as if politics too did not try to grapple with "facts") and relies on an odd definition of eugenics. For instance, he argues that one could support a Norwegian sterilization law that caused "legally incompetent persons to be sterilized with consent of guardian if serious mental illness, or high degree retarded or enfeebled, and reason to believe they could not support themselves or their offspring, or condition transferred to offspring" and not support eugenics ("Progress of Eugenics," 317). But what was eugenics if not the sterilization of those that could not support one's offspring, or of those with a condition that would detrimentally affect society as a whole? It was after all, the means by which the "fit" were to reproduce more than the "unfit."

125. Lily Kay, *The Molecular Vision of Life: Caltech, the Rockefeller Foundation, and the Rise of the New Biology* (Oxford: Oxford University Press, 1993), 8.

126. Ibid., 9.

127. Ibid., 17.

128. Ibid., 49.

129. Ibid., 92.

130. Ibid., 282.

131. Kay, *Who Wrote the Book of Life?*, xvi.

132. Ibid., 3.

133. Jacob, *Logic of Life*, 251.

134. Foucault, *Birth of Biopolitics*, 63–64.

135. Claire J. Tomlin and Jeffrey D. Alexrod, *Nature Reviews Genetics* 8 (May 2007): 331.

136. I. C. Weaver, N. Cervoni, F. A. Champagne, A. C. D'Alessio, S. Sharma, J. R. Seckl, S. Dymov, M. Szyf, and M. J. Meaney, "Epigenetic Programming by Maternal Behavior," *Nature Neuroscience* 7 (2004): 847–854.

137. Personal correspondence, September 18, 2009.

138. Catherine Malabou, *What Should We Do with Our Brain?*, trans. Sebastian Rand (New York: Fordham University Press, 2008), 41, 50; emphasis in original.

139. Ibid., 72.

The Undead of Information

1. See Matthew Kirschenbaum, *Mechanisms: New Media and the Forensic Imagination* (Cambridge, Mass.: MIT Press, 2008).

2. Sigmund Freud, "A Note Upon the 'Mystic Writing-Pad,'" in *The Standard Edition of the Complete Psychological Works of Sigmund Freud*, volume XIX (1923–1925): *The Ego and the Id and Other Works* (New York: W. W. Norton & Company, 1990), 230.

3. Jacques Derrida, *Archive Fever: A Freudian Impression*, trans. Eric Prenowitz (Chicago: University of Chicago Press, 1996), 19, 91–92.

4. Ibid., 15.

5. Warren J. Kurse II and Jay G. Heiser, *Computer Forensics: Incident Response Essentials* (Boston: Addison-Wesley, 2002), 77.

6. François Jacob, *Of Flies, Mice, and Men*, trans. Giselle Weiss (Cambridge, Mass.: Harvard University Press, 1998), 20–21. See Walter Gilbert, "The RNA World," *Nature* 319 (February 1986): 618.

7. Jacques Derrida, *Of Grammatology*, corrected ed., trans. Gayatri Chakravorty Spivak (Baltimore, Md.: Johns Hopkins University Press, 1997), 9.

8. Karl Marx, *Capital Volume One*, trans. Ben Fowkes (New York: Vintage, 1977), 163.

9. Thomas Keenan, "The Point Is to (Ex)Change It: Reading *Capital* Rhetorically," in *Fetishism as Cultural Discourse*, ed. Emily Apter and William Pietz (Ithaca, N.Y.: Cornell University Press, 1993), 165, 168.

4 Always Already There, or Software as Memory

1. Jutta Read-Scott, "Preserving Research Collections: A Collaboration between Librarians and Scholars," published by the Association of Research Libraries and the Modern Language Association, and the American Historical Association on behalf of the Task Force on the Preservation of the Artifact at the Annual Meeting of the Modern Languages Association, Chicago, December 27–30, 1999, <http://www.mla.org/rep_preserving_collections>, accessed 8/8/2010.

2. Ibid., "Introduction."

3. Abby Smith, *Why Digitize* (Washington, D.C.: Council on Library Resources, 1999): 3, 5, <http://www.eric.ed.gov:80/ERICDocs/data/ericdocs2sql/content_storage_01/0000019b/80/17/5d/23.pdf>, accessed 9/1/2009.

4. Archive as "always already there" is evident in Reginald Punnett's and Charles Davenport's texts discussed in chapter 3, in which the accumulation of human knowledge drives scientific and "civilized" progress. In its more dynamic form, software has also come to ground the "solution" to the traditional archive, the difficulties of access and selection outlined by Vannevar Bush in "As We May Think": hypertext as memex erases questions of reading as it focuses our attention on mapping connections. Even in its most critical form, Michel Foucault's formulation of the archive—not as something accumulated, but rather as first the law of what can be said—resonates with the logic of software. Discourse follows certain rules, its "networks" making possible certain positions of enunciation ("users").

5. See Geoffrey C. Bowker, *Memory Practices in the Sciences* (Cambridge, Mass.: MIT Press, 2005), 100.

6. John von Neumann, "First Draft of a Report on the EDVAC," Contract No. W-670-ORD-4926 between the United States Army Ordnance Department and the University of Pennsylvania, June 30, 1945, <http://qss.stanford.edu/~godfrey/vonNeumann/vnedvac.pdf>, 9, accessed 9/7/2009.

7. Von Neumann states, it is "only a transient standpoint, to make the present preliminary discussion possible. After the conclusions of the preliminary discussion, the elements will have to be reconsidered in their true electromagnetic nature. But at that time the decisions of the preliminary discussion will be available, and the corresponding alternatives accordingly eliminated" (Ibid.).

8. See William Aspray, *John von Neumann and the Origins of Modern Computing* (Cambridge, Mass.: MIT Press, 1990): 42.

9. Von Neumann, "First Draft," 3.

10. Warren S. McCulloch, "What Is a Number, that a Man May Know It, and a Man, that He May Know a Number?," in *Embodiments of Mind* (Cambridge, Mass.: MIT Press, 1965), 9.

11. Von Neumann, "Theory of Self-Reproducing Automata," *Papers of John von Neumann on Computers and Computer Theory*, Charles Babbage Institute Reprint Series for the History of Computing, vol. 12 (Cambridge, Mass.: MIT Press, 1986), 447.

12. Von Neumann, "First Draft," section 4.2.

13. Claude Shannon, "A Symbolic Analysis of Relay and Switching Circuits" (1936), in *Claude Elwood Shannon, Collected Papers*, ed. N. J. A. Sloane and Aaron D. Wyner (New York: IEEE Press, 1993).

14. Ibid., 2.

15. Von Neumann, "General and Logical Theory of Automata," *Papers of John von Neumann*, 396. Jay Forrester allegedly abandoned his initial plan to create an analog universal flight simulator precisely because the complex nature of the machine made distinguishing between noise and signal difficult.

16. Ibid., 398.

17. Ibid., 400–401.

18. Thomas D. Truitt and A. E. Rogers, *Basics of Analog Computers* (New York: Rider, 1960), 1–3; emphasis in original.

19. In the 1960s, analog computers were used in reactor engineering to solve problems related to "(1) automatic control rod drives; (2) stability of reactor power plants with internal feedbacks; (3) studies of reactivity lifetime reactor fuels; (4) dynamics of heat transfer and coolant flow" (Lawrence T. Bryant et al., *Introduction to Electronic Analogue Computing* [Chicago: Argonne National Labs, 1960], 9).

20. Lawrence T. Bryant and colleagues separate analyzers from analog computers based on the use of electronics (*Introduction to Electronic Analogue Computing*, 9). By 1948, the REAC (Reeves Analog Computer) had arrived. By 1949, Hartree noted the growing (wrongheaded) consensus in the United States to refer to (what would become) computers as analog machines and digital machines; Edmund C. Berkeley in his 1949 *Giant Brains, or Machines that Think* explained: "Machines that handle information as measurements of physical quantities are called *analogue* machines, because the measurement is *analogous* to, or like, the information" (New York: Wiley, 65). The fact that differential analyzers, along with early digital machines, were similarly dubbed "robot brains" or calculators and then became (retroactively) "computers" in the 1950s onward (the term *computer* makes far more sense in terms of numerical machines rather than differential analyzers, since both human and digital computers use numerical methods) reveals the extent to which analog and digital machines were considered to be analogous. The fact that differential analyzers would become "analogy" machines, and analog would constantly be confused with continuous, reveals the details of their coemergence and codependence.

21. U.S. War Department, "F U T U R E" press release, Bureau of Public Relations, Saturday, February 16, 1946, <http://americanhistory.si.edu/collections/comphist/pr2.pdf>, accessed 9/7/2009. The press release does go on to explain the difference between "digital" (general purpose) and "continuous signal" (specialized).

22. Vannevar Bush and Samuel H. Caldwell, "A New Type of Differential Analyzer," *Journal of the Franklin Institute*, 240 (1945): 255.

23. Ibid., 262.

24. John J. Rowlands, Director of News Service, MIT, press release, Tuesday, October 30, 1945.

25. Ibid.

26. Ibid.

27. Martha G. Morrow, "Magic Brains Spur Science and Technology: Fabulous Electronic Computers Herald Faster Planes and More Accurate Guns," in *New York World-Telegram*, March 10, 1946.

28. Herbert B. Nichols, "New Mathematical Robots Unscramble Digits to Multiply Inventions: Research Labs Calculate Devices to Bridge Years of Two Plus Two," in *Christian Science Monitor*, March 20, 1946, 11.

29. Larry Owens, "Vannevar Bush and the Differential Analyzer," in *From Memex to Hypertext: Vannevar Bush and the Mind's Machine*, ed. James M. Nyce and Paul Kahn (Boston: Academic Press, 1991), 23.

30. These resemblances are not merely accidental or natural, since charge was conceived in terms of force and flow, and each concept is used to elucidate the others.

31. As quoted by Larry Owens, "Vannevar Bush and the Differential Analyzer: The Text and Context of an Early Computer," *Technology and Culture* 27, no. 1 (January 1986): 66.

32. James M. Nyce, "Nature's Machine: Mimesis, the Analog Computer, and the Rhetoric of Technology," in *Computing with Biological Metaphors*, ed. Ray Paton (London: Chapman & Hall, 1994), 417. Differential analyzers "did not reduce phenomena to any set of general principles and then treat those principles as the thing itself. In effect then, they both follow and obey the same laws . . . these machines represent a triumph over a tendency in science to formalize and typify phenomena" (418).

33. Hartree, *Calculating Instruments and Machines*, 1.

34. Nyce, "Nature's Machine," 420. Furthermore, although analog machines (as Nyce later argues) held out "a possibility that for the brain or the machine to be understood, neither has to be reduced to any one set of laws or logical principles" (422), differential analyzers and other computing machines were called a "roomful of brains" in the popular press and in MIT's own press releases. The word brain, rather than mind, highlights physical particularity, and the description of these machines as brains draws on similarities between mechanisms/inputs: electricity is like blood or food, which gives the machine energy; the gears are like cells; and the differential analyzer has only one mechanical thought: differential equations. The robot brain thus operated by imitation, rather than duplication.

35. Indeed, it is strange and revealing that many scholars who argue that analog and digital computers thrived together during the 1940s to 1960s nonetheless base their analyses solely on the differential analyzer.

36. Bush and Caldwell, "New Type of Differential Analyzer," 277.

37. Operational amplifiers, as their name implies, amplify voltage differences. The simplest op-amp, a triode amplifier, comprises an electron tube with three electrodes, a cathode, a plate, and a control grid. As Truitt and Rogers explain, the cathode is heated to a high temperature by an electrical hot-wire filament, a positive voltage is applied between the plate and the cathode, and electrons are ejected and directed to the plate. The number of electrons per second traveling to the plate, namely, the plate current, depends on the temperature of the cathode and the applied voltage. If a negative voltage is applied between the control grid (located between the cathode and the plate) and the cathode, then the value of this grid voltage exercises far greater control over the plate current than does the value of the plate voltage. A small change in grid voltage can thus cause a relatively large change in plate current (Truitt and Rogers, *Basics of Analog Computers*, 59).

Because they enable relatively low-cost signal amplification, op-amps were key to long-distance telephone communications. Their amplification, however, is noisy and nonuniform: they usually amplify a rapidly changing voltage less than a steady voltage, and the amplification is time delayed (instead of giving an immediate amplification, in other words, there is more of a curve). Negative feedback, which Harold Black "discovered" in 1927, makes op-amps less noisy and more like an on-off switch. As David Mindell argues, what was key to "discovering" negative feedback was conceptualizing the "output . . . as containing a pure, desirable component—the signal—and an impure, unwanted component—the distortion" (*Between Human and Machine*, 118).

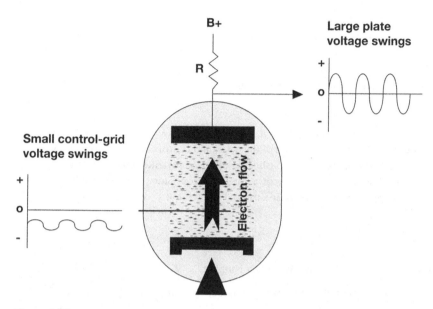

Figure 4.12
Basic triode

38. B. Holbrook and Uta C. Merzbach, "Interview," May 10, 1969, *Computer Oral History Collection, 1969–1973, 1977, Archives Center, National Museum of American History, Smithsonian,* <http://invention.smithsonian.org/downloads/fa_cohc_tr_holb_690510.pdf>,10, accessed 8/8/2010.

39. War Department, "F U T U R E," 3.

40. Hartree, *Calculating Instruments and Machines,* 15.

41. Ibid., 111. In a historical irony, Hartree's division between programming and coding, designed in part to separate mere coders from mathematician programmers, has disappeared, as has, arguably, the act of programming itself.

42. McCulloch, "What Is a Number," 1.

43. "Because of the 'all-or-none' character of nervous activity," they write, "neural events and the relations among them can be treated by means of propositional logic" (Walter McCulloch and Walter Pitts, "A Logical Calculus of the Ideas Immanent in Nervous Activity," in *Embodiments of Mind,* 19). They also state, "Many years ago one of us, by considerations impertinent to this argument, was led to conceive of the response of any neuron as factually equivalent to a proposition which proposed its adequate stimulus. He therefore attempted to record the behavior of complicated nets in the notation of the symbolic logic of proposition. The 'all-or-none' law of nervous activity is sufficient to ensure that the activity of any neuron may be represented as a proposition. Physiological relations existing among nervous activities correspond, of course, to

relations existing among the propositions; and the utility of the representation depends upon the identity of these relations with those of the logic of propositions" (21).

44. Ibid., 3.

45. McCulloch and Pitts, "A Logical Calculus," 38.

46. Ibid., 22.

47. McCulloch, "What Is a Number," 8. In "A Logical Calculus," McCulloch and Pitts write: "To psychology, however, defined, specification of the net would contribute all that could be achieved in that field—even if the analysis were pushed to ultimate psychic units or 'psychons,' for a psychon can be no less than the activity of a single neuron. Since that activity is inherently propositional, all psychic events have an intentional, or 'semiotic,' character" (38).

48. Ibid., 24.

49. Ibid., 22.

50. Russell as quoted by McCulloch, "What Is a Number," 6–7.

51. Warren McCulloch, "Why the Mind Is in the Head," in *Embodiments of Mind*, 73.

52. Ibid.

53. Seymour Papert, "Introduction," in *Embodiments of Mind*, xvi.

54. J. Y. Lettvin, H. R. Maturana, W. S. McCulloch, and W. H. Pitts, "What the Frog's Eye Tells the Frog's Brain," in *Embodiments of Mind*, 251.

55. McCulloch and Pitts, "A Logical Calculus," 35. More strongly, McCulloch would state in 1951 that the "cortex computes" ("Why the Mind Is in the Head," 85).

56. Turing's logic itself remarkably depends on analogies, which become equivalences.

57. McCulloch, "Why the Mind Is in the Head," 72.

58. Ibid., 91.

59. McCulloch and Pitts, "A Logical Calculus," 35.

60. Ibid., 35.

61. Bowker, *Memory Practices in the Sciences*, 100.

62. Warren S. McCulloch, "Finality and Form," in *Embodiments of Mind*, 275.

63. McCulloch, "Why the Mind Is in the Head," 84.

64. Ibid.

65. Ibid.

66. For more on this see Paul Edwards, *The Closed World: Computers and the Politics of Discourse in Cold War America* (Cambridge, Mass.: MIT Press, 1996).

67. McCulloch and Pitts, "A Logical Calculus," 38.

68. Warren S. McCulloch, "Physiological Processes Underlying Psychoneuroses," in *Embodiments of Mind*, 373.

69. Warren S. McCulloch, "Toward Some Circuitry of Ethical Robots or an Observational Science of the Genesis of Social Evaluation in the Mind-Like Behavior of Artifacts," in *Embodiments of Mind*, 197.

70. John von Neumann, *Theory of Self-Reproducing Automata* (Urbana: University of Illinois, 1966), 39.

71. Ibid., 40.

72. John von Neumann, *The Computer and the Brain*, 2nd ed. (New Haven, Conn.: Yale University Press, 2000), 61.

73. McCulloch, "Why the Mind Is in the Head," in *Embodiments of Mind*, 102. Current studies of memory have moved away from this stark distinction between order and data back toward an analogy-based system of memory that insists that the brain preserves traces of events through the combination of neurons.

74. Von Neumann, *Theory of Self-Reproducing Automata*, 39.

75. Jacques Derrida, *Archive Fever: A Freudian Impression* (Chicago: University of Chicago Press, 1998), 19; emphasis in original.

76. Arthur W. Burks, Herman H. Goldstine, and John von Neumann, "Preliminary Discussion of the Logical Design of an Electronic Computing Instrument," Part I, Volume 1 (Princeton, N.J.: Institute for Advanced Study, 1946).

77. Von Neumann, *An Electronic Computing Instrument*, section 4.4 (6).

78. Marshall McLuhan, *Understanding Media: The Extensions of Man* (Cambridge, Mass.: MIT Press, 1964).

79. Von Neumann, *The Computer and the Brain*, 36.

80. McCulloch, "Why the Mind Is in the Head," 133.

81. As quoted in McCulloch, "Why the Mind Is in the Head," 92–93. Von Neumann's move to files is intriguing, especially if one considers the importance of files—and of disposing files—to modern bureaucracy and state power (see Cornelia Vismann's *Files: Law and Media Technology*, trans. Geoffrey Winthrop-Young [Stanford, Calif.: Stanford University Press, 2008]).

82. As Derrida notes, a trace that cannot fade is not a trace but a full presence, the son of God (Jacques Derrida, "Freud and the Scene of Writing," *Writing and Difference*, trans. Alan Bass [Chicago: University of Chicago Press, 1978], 230).

83. Von Neumann, *The Computer and the Brain*, 65.

84. For more on the relationship between neoliberalism and Nazism, see Michel Foucault, *The Birth of Biopolitics: Lectures at the Collège de France, 1978–1979*, trans. Graham Burchell (Basingstoke, England, and New York: Palgrave Macmillan, 2008), 106–116.

85. William Poundstone, *Prisoner's Dilemma* (New York: Doubleday, 1992), 194.

86. Nicholas Vonneuman, "The Philosophical Legacy of John von Neumann, in the Light of Its Inception and Evolution in His Formative Years," paper presented at The Legacy of John von Neumann Symposium at Hoftstra University, May 30, 1988 (Meadowbrook, Pa.: N. A. Vonneuman, 1987–1988), 1.

87. Ibid., 2.

88. Ibid.

89. Johannes Wolfgang Goethe, "Faust's Study (i)," *Faust Part One*, trans. David Luke (Oxford: Oxford University Press, 1987), lines 1225–1237.

90. Von Neumann, "General and Logical Theory," 392.

91. Von Neumann writes, "I will introduce as elementary units neurons, a 'muscle,' entities which make and cut fixed contacts, and entities which supply energy, all defined with about that degree of superficiality with which the formal theory of McCulloch and Pitts describes an actual neuron by axiomatizing automata in this manner, one has thrown half of the problem out the window, and it may be the more important half. One has resigned oneself not to explain how these parts are made up of real things, specifically, how these parts are made up of actual elementary particles, or even of higher chemical molecules" ("Theory of Self-Reproducing Automata," 480).

92. Von Neumann, "General and Logical Theory," 412–413.

93. Ibid., 413. Von Neumann also offers an interpretation of Turing's "On Computable Numbers, with an Application to the Entscheidungsproblem" (see note 117) in order to justify this statement. Von Neumann writes:

Turing observed that a completely general description of any conceivable automaton can be (in the sense of the foregoing definition) given in a finite number of words. The description will contain certain empty passages—those referring to the functions mentioned earlier . . . which specify the actual functioning of the automaton. When these empty passages are filled in, we deal with a specific automaton. As long as they are left empty, this schema represents the general definition of the general automaton. Now it becomes possible to describe an automaton, which has the ability to interpret such a definition. In other words, which, when fed the functions that in the sense described above define specific automaton, will thereupon function like the object described. The ability to do this is no more mysterious than the ability to read a dictionary and a grammar and to follow their instructions about the uses and principles of combinations of words. This automaton, which is constructed to read a description and to imitate the object described is then the universal automaton in the sense of Turing. (Von Neumann, "General and Logical Theory," 416)

94. Von Neumann, "Theory and Organization of Complicated Automata," *Papers of John von Neumann*, 450.

95. Von Neumann, "General and Logical Theory," 414.

96. Arthur Burks, "Introduction: Theory of Natural and Artificial Automata," *Papers of John von Neumann*, 374.

97. Von Neumann, "General and Logical Theory," 419.

98. Ibid.

99. Ibid., 420.

100. Ibid.

101. Ibid., 421.

102. Von Neumann, *Theory of Self-Reproducing Automata*, 101 and 113. In other models of self-reproduction, such as the cellular model, von Neumann still privileged the role of memory. It is only with a large external yet accessible memory that his cellular units can be logically universal—capable of inductive processes—and thus function as fundamental cells. In describing the function of the construction of a "tape and its control" for the cellular automata, von Neumann treats the memory as containing instructions for the creation of a secondary automata (Ibid., 202).

103. Von Neumann, *Papers of John von Neumann*, 368. This comparison, however, occurs not only on the level of mathematics or mathematization, but also on the level of heuristics, descriptions, and strategies.

104. John von Neumann and Oskar Morgenstern, *Theory of Games and Economic Behavior* (Princeton, N.J.: Princeton University Press, 1947), 9.

105. Ibid., 7.

106. Ibid., 48.

107. Ibid., 79.

108. Ibid., 49.

109. As quoted in Poundstone's *Prisoner's Dilemma*, 168.

110. Von Neumann and Morgenstern, *Theory of Games*, 31.

111. Milton Friedman, *Capitalism and Freedom*, Fortieth Anniversary Edition (Chicago: Chicago University Press, 2002), 25.

112. A. M. Turing, "On Computable Numbers, with an Application to the Entscheidungsproblem," *Proceedings of the London Mathematical Society* 2, no. 42 (1937): 230–265.

113. Ibid., section 6.

114. Von Neumann, *The Computer and the Brain*, 70–71.

115. Ibid., 71.

116. Ibid., 72–73; emphasis in original.

117. Frances A. Yates, *The Art of Memory* (Chicago: University of Chicago Press, 2001), 6–7.

118. See Michael R. Williams, *A History of Computing Technology* (Los Alamitos, Calif.: IEEE Computer Society Press, 1977), 306–316.

119. Derrida, "Freud and the Scene of Writing," 226.

120. Wolfgang Ernst, "Dis/continuities: Does the Archive Become Metaphorical in Multi-Media Space," in *New Media Old Media*, ed. Wendy Hui Kyong Chun and Thomas Keenan (New York: Routledge, 2006), 118. Although this is certainly true for CRT screens, it is not necessarily true for LCD screens, which operate more like blinds that allow certain sections of light through.

121. Matthew Kirschenbaum, *Mechanisms: New Media and the Forensic Imagination* (Cambridge, Mass.: MIT Press, 2008).

122. Repetition not only grounds the archive, it also threatens it. Drawing from Freud's work on the death drive, Derrida argues, "Repetition itself, the logic of repetition, indeed the repetition compulsion, remains . . . indissociable from . . . destruction" (*Archive Fever*, 11–12).

123. Jeffrey Toobin, "Scotus Watch," The Talk of the Town, in *The New Yorker*, November 11, 2006, <http://www.newyorker.com/talk/content/articles/051121ta_talk_toobin>, accessed 01/27/2006.

124. This is because there are no shelves, no fixed relation between what is storeable and the place they are stored. As Harriet Bradley has argued, the Internet breaks the bond between location and storage: if before "only what has been stored can be located," now "memory is no longer located in specific sites" ("The Seductions of the Archive: Voices Lost and Found," *History of the Human Sciences* 12, no. 2 [1999]: 113).

125. <http://www.archive.org/about/about.php>, accessed 2/1/2007.

126. <http://www.archive.org/about/about.php>, accessed 2/1/2007.

127. Ernst, "Dis/continuities," 119.

128. See Paul Virilio, "The Visual Crash," in *CTRL [SPACE]: Rhetorics of Surveillance from Bentham to Big Brother*, ed. Thomas Y. Levin et al. (Cambridge, Mass.: MIT Press, 2002), 108–113; *Open Sky*, trans. Julie Rose (London: Verso, 1997); and "Speed and Information: Cyberspace Alarm!," <http://www.ctheory.net/text_file.asp?pick=72>, accessed 2/1/2007.

Conclusion

1. Thomas Keenan, "The Point Is to (Ex)Change It: Reading *Capital* Rhetorically," in *Fetishism as Cultural Discourse*, ed. Emily Apter and William Pietz (Ithaca, N.Y.: Cornell University Press, 1993), 184–185.

2. Erwin Schrödinger, *What Is Life?* with *Mind and Matter* and *Autobiographical Sketches* (Cambridge, Mass.: Cambridge University Press, 1992), 21–22.

3. Ibid., 440.

4. Fredric Jameson, *Postmodernism, or The Cultural Logic of Late Capitalism* (Durham, N.C.: Duke University Press, 1991), 51.

5. For more on this see Michel de Certeau's *The Practice of Everyday Life,* trans. Steven Rendall (Berkeley: University of California Press, 1984).

6. This is the topic of my next book, "Imagined Networks."

7. Jacques Derrida stresses that writing represents the disappearance of the origin: "To repeat: the disappearance of the good-father-capital-sun is thus the precondition of discourse, taken this time as a moment and not as a principle of *generalized* writing. . . . The disappearance of truth as presence, the withdrawal of the present origin of presence, is the condition of all (manifestation of) truth. Nontruth is the truth. Nonpresence is presence. Differance, the disappearance of any originary presence, is *at once* the condition of possibility *and* the condition of impossibility of truth. At once" ("Plato's Pharmacy," *Dissemination,* trans. Barbara Johnson [Chicago: University of Chicago, 1981], 168).

8. Intriguingly, Slavoj Žižek links specters to a fear of freedom in his introduction to *Mapping Ideology* (London: Verso, 1994), 27.

9. Milton Friedman, *Capitalism and Freedom,* Fortieth Anniversary Edition (Chicago: Chicago University Press, 2002), lx. Michel Foucault discusses civil society as an umbrella term to bring together the subject of rights and the *homo oeconomicus* in *The Birth of Biopolitics: Lectures at the College de France, 1978–1979,* trans. Graham Burchell (Basingstoke, England, and New York: Palgrave Macmillan, 2008), 291–316.

10. See David Harvey, *A Brief History of Neoliberalism* (Oxford: Oxford University Press, 2005).

11. Vicente Rafael, "The Cell Phone and the Crowd: Messianic Politics in the Contemporary Philippines," *Public Culture* 15, no. 3 (2003): 419.

Epilogue

1. See Toni Morrison, *Playing in the Dark: Whiteness and the Literary Imagination* (New York: Vintage, 1993), 63.

2. For more on this, see Wendy Hui Kyong Chun and Lynne Joyrich, eds., *Race and/as Technology,* special issue of *Camera Obscura* 24 (2009), and Wendy Hui Kyong Chun, "Race as Archive," *Vectors: Journal of Culture and Technology in a Dynamic Vernacular* 3, no. 1 (2007).

Index

Abstraction, viii, 176
  biological, 140–157
  data abstraction, 37–39
  in Marx, 135
Aiken, Howard, 30
Allen, Garland, 122
Allen, George (senator), 98
Althusser, Louis, 66, 73
Antonelli, Mauchly Kathleen (McNulty), 29, 32
Archaeology, 112–114. *See also* Archive;
  Foucault, Michel
Archive, 97–100, 112–114
  and the digital, 3, 137–140, 170–172, 176,
    177 (*see also* Internet)
  erasure of, 212n18
  and genetics, 76–79, 104, 120–124
Aristotle, 99
  on metaphor, 56
Aspray, William, 34
Austin, J.L., 28
Automata, 10, 115–116, 140, 157–158,
  163–164, 228n91, 228n93, 229n102

Babbage, Charles, 7, 38, 55, 158
Backus, John, 42
Barthes, Roland, 15
Bartik, Jean Jennings, 29, 31, 44, 159
Bateson, Gregory, 165–166
Bateson, William, 117–118
Baudrillard, Jean, 90
  The Ecstasy of Communication, 16

Benington, Herbert D., 4, 35
Berners, Tim, 17
Bilas, Frances (Spence), 29
Biology. *See* Cybernetics; Eugenics;
  Genetics
  and computer technology, vii, 98,
    101–104, 112–116, 141
  and physics, 110–112 (*see also*
  Schrödinger, Erwin)
Blackboxing, 45, 60, 140, 141
Blanch, Gertrude, 38–40
Boas, Franz, 120
Broy, Manfred, 3, 19
Brown, Bill, 11
Brucker-Cohen, Jonah, 94
Burks, Arthur, 34, 163, 164
Bush, George W., 16
Bush, Vannevar, 98–99, 106, 110,
  144, 147, 151, 161, 170
  "As We May Think," 75–81, 116,
    208n66, 221n4
  memex, the, 75–80
Butler, Judith, 28

Caldwell, Samuel, 148, 151
Causality, 10, 63, 65, 67, 69, 103, 112,
  155–156, 179. *See also* Indexicality
Caygill, Howard, 99
Ceccarelli, Leah, 111
Ceruzzi, Paul, 3–4
Chu, Chuan J., 4, 33

Code, 113, 114, 165–166, 199n29
  as fetish, 28, 49–54
  genetic, 103–105, 107, 111, 112, 125,
    128–129, 176, 179, 215n40
  as *logos*, 19–55, 59–60, 89, 99, 103, 110, 112,
    116, 125, 135, 152–153, 166–167, 175–177
  pseudocode, 19, 41–42, 188n3, 197n115
  source code (*see* Sourcery)
  the Morse, 109, 111
Colin, Gordon, 7
Commodification, 9, 19, 41, 72. *See also*
  Code
Computation, 17, 19, 22, 37–38, 60, 67, 91,
    103, 129, 148, 157–159, 167, 177. *See also*
  Computer, the
  and commands, 30
Computer, the, 2, 7, 11, 31, 46, 49, 62, 68,
    80, 85–86, 90–92, 97, 102, 108, 112,
    133–134, 177, 199n129, 207n60, 211n4. *See
  also* Archive; Computation; ENIAC, the;
  Eugenics; Memory
  and agency, 5, 90
  analog, 24, 75, 81, 222n19, 223n20
  and biology, 104–105, 107, 114, 116, 120,
    129, 176–177
  and biopower, 9–10, 27–28, 128
  and cognitive psychology, 101–102
  computer games, 70
  computer networks, 90 (*see also* Maps,
    mapping)
  computer program, 47, 51, 67, 70, 107, 128
  digital, 22, 26, 30, 60, 75, 77, 81, 103, 104,
    142–144, 148, 152, 154, 161, 170, 223n20
  and ideology, 66
  and knowledge, 6
  and/as metaphor, 18, 55–59, 64, 82, 98, 102,
    157, 175
  women as, 4, 29, 33, 38, 39, 44, 192n46
Comrie, Leslie J., 39
Control, 6–8, 18, 28, 29, 32–33, 35–38,
    50–51, 60, 62, 66–67, 73, 83–91, 99,
    103–108, 115, 121, 130–138, 177. *See also*
  Cybernetics; Freedom; Power
  grid, 224n37

Corbato, Fernando, 88
Correns, Carl, 117
Crick, Francis, 107, 110, 111
Cybernetics, 6, 33, 60, 104–107, 122, 128,
    129, 152. *See also* Eugenics
  as memory, 135

Daemon, 55, 88–89, 210n93. *See also* UNIX,
  UNIX daemon
  daemonic interfaces, 59–97
  Maxwell's daemon, 88
Darwin, Charles, 117, 120
Davenport, Charles, 119, 126
  Heredity in Relation to Eugenics, 122–125
Dawkins, Richard, 102, 103
de Brosses, Charles, 50
Deleuze, Gilles, and Félix Guattari, 49,
    101
de Prony, Gaspard Clair François
  Marie Riche, 38
Derrida, Jacques, 28, 80, 97–99, 134, 135,
    169, 170, 190n25, 191n29, 208n77
  and Felix Guattari, 49, 101
Dibbell, Julian, 69
Differential analyzer, the, 144–152. *See also*
  Bush, Vannevar
Digital, the, 5, 9, 10, 26, 33, 95, 97–98, 115,
    133, 137–141, 154, 158–159, 169, 171–173.
  See also Computer, the
  digitization, 11, 137
  images, 15–16
Dijkstra, Edsger, 36
Direct manipulation, 59, 62–66, 69, 176
DNA, 2, 6, 92, 103, 105, 107, 111–112, 114,
    128, 134–135, 179
Doane, Mary Ann, 15, 187n4
Doyle, Richard, 105–106, 111

Eck, David, 37
Eckert, J. Presper, 33, 140, 157
Edwards, Paul N., 2, 30, 46, 55, 60, 64, 101,
    204n25
Enduring ephemeral, the, 10, 95, 128, 133,
    137–138, 167–173

Engelbart, Douglas, 59, 76, 80, 110, 176, 208n66
demo, 83–85
ENIAC, the, 4, 19, 29–34, 43, 46, 144, 152
girls, 29, 31, 33–34, 46, 192n46
"women of" (*see* ENIAC, the, girls)
Ernst, Wolfgang, 97, 100, 169, 171
Eugenics, 7, 38, 104, 116–129, 220n124. *See also* Biology
as nurture, 120–124
Everett, Bob, 42
Experian, 17

Facebook, xi, 13, 69–70, 98
Falk, Raphael, 118, 217–218n87
Fleming, Donald, 62, 222n15
Forrester, Jay, 62, 222n15
Foucault, Michel, 6, 8, 27, 29, 80, 112–114, 122–124, 129, 185n31, 185n32, 186n33, 208n74, 221n4. *See also* Archaeology; Governmentality
Freedom, xii, 7–8, 21, 43, 49, 51, 59, 62, 65–66, 74–75, 84, 94, 106, 129, 176–178, 185n32, 198n120, 208n66, 231n8
and control, 46, 185n32
Fuller, Matthew, vii–ix, 94
Future, the, xi–xii, 2, 8–9, 11, 13, 33, 50, 78–92, 97–99, 101, 110–111, 115, 117, 125, 127, 133, 138, 156, 167, 176, 186. *See also* Enduring ephemeral, the
and past, 54, 76, 171–172, 206n47

Galloway, Alexander, 21, 27, 33, 112
"Language Wants to Be Overlooked: On Software and Ideology," 22–24
Galton, Francis, 118–123
Gender, 90, 128
and programming, 4, 18, 29, 35, 46, 103, 176
Genetics, 2, 10, 80, 90, 103–107, 116–123, 126–127, 179, 214n28, 219n110. *See also* Biology; Code
breeding, 117–126, 179
and computing, xii, 9, 17

heredity, 79, 103–104, 107–108, 111–112, 115–128, 139, 157, 218n87, 219n124
Human Genome Project, 130
and software, 108–116
Gibson, William, 9, 75, 187n48
Goethe, Johann Wolfgang, 19, 161
Faust, 22, 27, 31, 161–164
Goldstine, Adele, 192
Goldstine, Herman, 3, 25, 34, 103, 191n27, 192n46
Good, I. J., 30
Google, xi, 13, 17, 92–93, 97, 137
Earth, 62 (*see also* Interface)
Governmentality, 6–9, 34, 109, 124, 127–129, 176, 185n31, 186n33
Grier, David, 38
Guttag, John V., 38
Guattari, Félix. *See* Deleuze, Gilles, and Félix Guattari

Haberman, Seth, 90
Hagen, Wolfgang, 4
Hardware, xii, 1–4, 9–11, 17, 20–26, 32–34, 37, 45–46, 49, 55, 66–68, 77–78, 92, 97, 101–107, 112, 115, 139, 152, 156, 164, 175, 179, 183n3, 197n116, 214n23. *See also* Computer, the; Software
Hartree, Douglass, 223n20, 225n41
Calculating Instruments and Machines, 152
Harwood, Graham, 52
Holberton, Frances Snyder, 29, 32–34, 44, 159
Hopper, Grace Murray, 4, 30–31, 33–34, 43, 45, 110, 112, 197n115, 198n120, 198n121
Hume, David, 154, 156
Huxley, Julian, 122

Ideology, 2, 17–18, 59, 66, 205n44. *See also* Interface
Indexicality, 15, 18, 68, 92, 179
Information, ix, 5–6, 11, 15–17, 21, 25, 37, 42, 45, 49, 53, 57, 71, 74, 77–80, 87, 90–94, 97–98, 101, 104–111, 116, 127–128, 137–142, 151–173, 185n28, 197n115, 198n75, 212n18, 214n21, 223n20. *See also* Maps

Information (cont.)

Government Information Awareness (GIA),
93

Information Awareness Office (IAO),
16

the undead of, 133–135

Total Information Agency (ITA), 90

"total information" systems, 15

networks, 74

Interactivity, 61–62, 71, 89, 93

Interface, xii, 2, 8–10, 17–18, 22, 28, 33, 37,
49, 53, 55, 114, 175–177, 204n20, 205n44,
206n47. *See also* Maps; Metaphor

as ideology, 66–68

daemonic, 59–97

GUI (Graphical User Interface), 8, 59–60, 66,
176, 206n47

WYSIWG (What You See Is What You Get),
18

Internet, xi, 6, 17, 52, 62–80, 91, 97, 138,
170–171, 212n18, 230n124

Internet Wayback Machine (IWM), 138,
170–171

Iterability, 25, 28, 47, 191n29

Jacob, François, 103–115, 125–128

The Logic of Life, 125

Jameson, Fredric, 71–76, 206n47, 207n53

Johannsen, Wilhelm, 119–120

Johnson, Mark. *See* Lakoff, George, and Mark
Johnson

Kay, Lily, 104–105, 110, 115–117, 127–128,
215n42

*Who Wrote The Book of Life? A History of the
Genetic Code*, 105, 127–128

Keller, Evelyn Fox, 111, 121

Kevles, Daniel J., 118–119

Kirschenbaum, Matthew, 5, 133, 170,
205n44

*Mechanisms: New Media and the Forensic
Imagination*, 170, 205n44

Kittler, Friedrich, 3, 92

Knowledge, xiii, 6–9, 15–22, 37, 43, 48–53,
55, 57, 72–75, 79–80, 89, 98, 108,110–115,
121–122, 137, 154–156, 175–180, 186n33,
195n75, 200n136, 221n4. *See also* Archive

knowledge-power, 21

visual, 33, 92, 180

Koss, Adele Mildred, 37, 41–42

Kraft, Philip, 35–36, 42

Lakoff, George, and Mark Johnson, 56–57, 63

Lalande, Joseph, 38

Landecker, Hannah, 130

Lapaute, Nicole-Reine, 38

Lat, David, 170

Laurel, Brenda

Computer as Theatre, 64–66, 204n20

Laplace, Pierre-Simon, 9, 101–109, 114–116,
176

Licklider, J. C. R., 62, 82

Lovink, Geert. *See* Theory, vapor

Lyotard, Jean-François, 72, 206n47

Machine, 17, 19–20, 55, 60, 97, 135, 195n80,
207n60, 223n20, 224n34, 211n4. *See also*
Computer, the; Turing, Alan

Mackenzie, Adrian, xii, 3

Mahoney, Michael, 3–4, 25, 36

Manovich, Lev, 69, 71, 206n47

The Language of New Media, 20, 91

Maps, 8, 20, 28, 59, 85, 175–177. *See also*
Archaeology; Ideology; Interface

cognitive, 18, 71–79, 207n51

mapping, 8–10, 51, 53, 58, 59–62, 69–70,
89–95, 113, 123, 129–130, 205n44,
221n4

of power, 28

Marx, Karl, 50, 52, 71, 116, 135, 175

Matrix, The (film), 66

Matrix, the (Multistate Anti-TerRorism
Information eXchange), 90

Mauchly, John, 34, 140

Mauchly, Mary, 192n46

McCarthy, John, 61

McCulloch, Warren, 26, 103, 141, 153–166,
  225n43, 226n47, 227n73, 227n81,
  228n91
McLuhan, Marshall, 159
McNulty, Kathleen. *See* Antonelli, Mauchly
  Kathleen (McNulty)
McPherson, Tara, 68–69
Memex, the. *See* Bush, Vannevar
Memory, ix, 4–11, 25, 55, 95–99, 175–177. *See
  also* Archive; Computer, the; Software
computer and/as, xii, 22–27, 41, 78, 80,
  89, 95, 101–107, 115–116, 120, 133–135,
  137–175, 194n75, 208n66, 208n77,
  211n4, 212n18, 214n28, 229n102,
  230n124
memory card, 15
Mendel, Gregor, 10, 79–80, 103–125, 179
Metaphor, viii, 2, 10, 17–18, 55–57, 59, 60,
  63–64, 72, 89–90, 94, 97, 100, 156, 162,
  202n13, 202n16, 204n20, 206n47. *See also*
  Computer, the; Software
Mez, 52
Microsoft, 21, 91, 94, 188n13
Mind, the, 2, 6, 22, 35–38, 50–53, 55, 62, 73,
  78, 83, 101, 154–168
Mindell, David, 75, 207n60, 225n37
MIT AI lab 46, 49
Moglen, Eben, 67
Morgan, T. H., 126
Morgenstern, Oskar, 165
Morrison, Toni, 179
Muller, Hermann J., 116–119, 217n87

Nelson, Ted, 76, 208n66
Neoliberalism, 7–10, 27, 29, 74, 129, 131,
  139, 161, 166, 178. *See also* Postmodernism
Neumann, John von, 3, 10, 25–27, 34, 55, 62,
  102–107, 112–116, 139–167, 190n17,
  211n4, 222n7, 222n15, 227n76, 228n86,
  228n93, 229n102
Neuromancer. See Gibson, William
Neuron, 102–103, 130, 139–162, 225n43,
  226n47, 228n91

New media, xi–xii, 1–2, 8, 11, 15, 20, 21, 75,
  91, 97, 100, 105, 159, 170, 172
new media studies, field of, 1, 20–21, 75, 91,
  205n44
Nietzsche, Friedrich, 89, 94
Nostalgia, 28, 75
Nyce, Paul, 148, 224n34

Owens, Larry, 145

Papert, Seymour, 155
Paranoia, 16, 53, 66, 200n144
Paul, Diane, 121, 126
Pauling, Linus, 110
Pearson, Karl, 118–119
Performative, the, 22, 27–28, 51, 175
Photography, 15
  photograph, 7, 15, 16, 62, 69
Piaget, Jean, 63
Pietz, William, 50, 52
Pitts, Walter, 26, 103, 141, 153–158, 162–166
Plant, Sadie, 33
Postmodernism, 10, 72–74, 206n47
Poundstone, William, 161
Power, xii, 1–10, 16–18, 19–21, 27–97, 100,
  106, 110–114, 170, 175–178, 186n33,
  196n89, 199n32, 227n81. *See also* Control;
  Maps
biopower, 103, 122–124, 128–129 (*see also*
  Cybernetics; Eugenics; Genetics)
Programming, xi–xii, 1–18, 25–62, 101, 130,
  152, 165, 175–177, 184n16, 191n29,
  195n80, 197n115, 199n132, 200n136,
  220n124, 225n41. *See also* Genetics
and pleasure, 110
direct programming, 19–20, 41
executability, 22, 27, 48, 190n14 (*see also*
  Code)
FORTRAN, 32, 42–45, 51, 104
programmability, 9–10, 49–54, 91, 98,
  101–104, 112–113, 116–117, 124–130, 156,
  173, 175
structured, 25, 36–37, 188n13, 196n89

Provine, William, 120, 127, 219n124
Punnett, Reginald, 118, 121–122, 221n4

Race, 50, 99, 119, 120, 123, 124, 126–127,
    179–180, 217n82, 218n91
Radin, Mary Jane, 4–6
RAND Corporation, the, 35
Rand, Remington, 45
Real, the, 15, 72, 199n132, 206
    real-time, 59, 60, 68, 81, 89, 172
Repetition, 25, 46, 52, 78–80, 98–99, 103,
    126, 133, 157, 170–173, 176, 230n122
Rhodes, Ida, 38–40
Rolls-Hansen, Nils, 119
Russell, Bertrand, 154

SAGE (Semi-Automatic Ground
    Environment), 4, 35, 60–62, 68
Sammet, Jean E., 44–45
Sanger, Margaret, 123
Sapp, Jan, 80, 117, 118, 124
Sarkar, Sahotra, 118
Saxe, John Godfrey, 1
Schrödinger, Erwin, 10, 125, 176, 215n40
    What Is Life?, 10, 103–112, 115–116
Sedgwick, Eve Kosofsky, 121
Seisint, 15
Sexuality, 33, 122–123, 185n31, 194n68
Seysenegg, Erich Tschermak von, 117
Shannon, Claude, 106–107, 109, 141–142,
    157
Shneiderman, Ben, 8, 59, 62, 83
Simpson, O. J., 16
Smith, Abby, 138
Smith, Adam, 38
Smith, Linda C., 75
Software, 1–13, 18, 175–180. See also
    Freedom; Ideology; Memory; Metaphor
    and ideology, 71–72
    and/as metaphor, 90–175
    as thing, xii, 6–11, 20–58
Source, 80, 89–99, 105–114, 118, 125, 128,
    137, 148, 177

Sourcery
    real-time, 68–72, 175
    source code, 5, 9–10, 19–55, 59, 135, 167,
        177, 188n13
Stallman, Richard, 49, 51
Stent, Gunther, 111
Stephenson, Neal, 30
Stern, Curt, 122
Stocking, George, Jr., 121
Subjectivity, 9, 18, 113
Surveillance, 13, 17, 58, 90, 92–93, 100
Systems Development Corporation (SDC), 35

Teitelbaum, Ruth Lichterman, 29
Theory, 20, 28, 54, 65, 71, 79, 82, 104,
    115–120, 130, 134, 140, 155. See also
    Neoliberalism
    game, 164–166
    vapor, 20–21, 189n7
Thing(s), xii, 1–13, 18, 28, 32, 41, 50–58, 74,
    97–99, 133, 135, 143–177, 228n91. See also
    Code; Software
Torvalds, Linus, 49
Transparency, 16–17, 54, 89, 92–95
Tscherma, Erich, 117
Turing, Alan, 30, 33, 41, 53–55, 112, 114–115,
    166–167, 176, 228n93
    Turing machine, viii, 103, 151, 155, 166

Ullman, Ellen, 52
UNIVAC (the Universal Automatic
    Computer 1), 33, 41, 42, 188n3,
    195n75, 197n115, 197n116
UNIX, 67, 87–89
    UNIX daemon, 55

Virilio, Paul, 172
Vismann, Cornelia, 85–86, 160
Vries, Hugo de, 117

Watson, James, 107, 110–111
Weaver, Warren, 106, 130
Weblog, 172

Weismann, August, 121

Weizenbaum, Joseph, 2, 17–18, 28, 46–51, 55–56, 65, 165, 176, 187n50, 199n136

Weldon, Walter Raphael, 118

Wescoff, Marlyn (Meltzer), 29

Whirlwind group, the, 35, 42

Wiener, Norbert, 46, 106, 109, 155, 214n21, 214n23

  Cybernetics, or, Control and Communication in the Animal and the Machine, 108

Work Progress Administration's (WPA) Math Tables Project (MTP), the, 38

Wrens, the, 30

Yates, Frances A., 168

Žižek, Slavoj, 52, 67

Zuse, Konrad, 104

Printed in the United States
by Baker & Taylor Publisher Services